Essential Microelectronic Circuits (Second Edition)

A student's guide

Online at: https://doi.org/10.1088/978-0-7503-5512-4

Essential Microelectronic Circuits (Second Edition)

A student's guide

Yumin Zhang
Southeast Missouri State University, Cape Girardeau, Missouri, USA

IOP Publishing, Bristol, UK

ISBN 978-0-7503-5512-4 (ebook)
ISBN 978-0-7503-5510-0 (print)
ISBN 978-0-7503-5513-1 (myPrint)
ISBN 978-0-7503-5511-7 (mobi)

DOI 10.1088/978-0-7503-5512-4

Version: 20251101

IOP ebooks

British Library Cataloguing-in-Publication Data: A catalogue record for this book is available from the British Library.

Published by IOP Publishing, wholly owned by The Institute of Physics, London

IOP Publishing, No.2 The Distillery, Glassfields, Avon Street, Bristol, BS2 0GR, UK

US Office: IOP Publishing, Inc., 190 North Independence Mall West, Suite 601, Philadelphia, PA 19106, USA

To my parents:

H Z Mao and S T Zhang

Contents

Preface

Microelectronics is often perceived by students as a particularly challenging subject—sometimes even described as confusing or disorganized. This perception arises from several factors. First, the field involves a wide range of electronic devices, including various types of diodes, transistors, and operational amplifiers, each with distinct properties and behaviors. Second, the characteristics of these devices are inherently complex—for instance, the nonlinear I–V curves of transistors are not always intuitive. Third, circuit configurations vary widely, such as common-base (CB), common-collector (CC), and common-emitter (CE) bipolar junction transistor (BJT) amplifiers, each exhibiting different performance attributes. Fourth, the interaction among components—especially in circuits with feedback—often introduces strong coupling effects that are difficult to analyze in isolation. Finally, circuit design requires balancing numerous and often conflicting constraints, such as bandwidth, power consumption, linearity, and stability. As a result, microelectronics is sometimes viewed as an esoteric skill, where mastery seems elusive without both theoretical insight and practical experience.

Before taking this course, most students have completed subjects such as *Circuit Analysis*, which offers a structured problem-solving approach grounded in Kirchhoff's laws. These courses emphasize well-defined principles that support the systematic application of equations. In contrast, microelectronics presents a different challenge. Its foundational concepts often lack the same level of mathematical rigor and must be understood through intuition, pattern recognition, and physical insight. Unfortunately, these less tangible aspects are frequently overlooked. A common pitfall is the overemphasis on memorizing equations without grasping the underlying principles. Given the vast number of formulas tied to various circuits, rote memorization is neither practical nor effective. This book seeks to bridge that gap by focusing on the general principles that govern circuit behavior, helping students build the intuition needed to derive and apply equations meaningfully in electronic circuit analysis and design.

The first edition of this book—*The Tao of Microelectronics*—was published in 2014, emphasizing conceptual understanding and key ideas in electronic circuits. While that approach proved helpful, it became clear that many students would benefit from more concrete examples. To meet that need, the second edition integrates a wide range of circuit simulations alongside theoretical explanations. These simulations serve to reinforce understanding and provide valuable insight into real-world circuit behaviors.

In addition to this expanded practical focus, the second edition introduces many new topics and delves deeper into existing ones. For example, the final chapter now presents a broad spectrum of oscillator circuits, encompassing both classical and modern designs. New analytical tools—such as frequency spectrum analysis and total harmonic distortion (THD) evaluation—are also introduced to help students better understand and evaluate circuit performance. These additions have significantly extended the scope of the book and offered a more comprehensive and application-oriented resource.

The first chapter begins with the fundamentals of circuit analysis, establishing a foundation in passive components, signal behavior, and frequency response. It then explores semiconductor devices in chapter 2, covering carrier dynamics and device operation—including diodes, BJTs, and metal–oxide–semiconductor field-effect transistors (MOSFETs). From there, the focus shifts to amplifier circuits in chapter 3, which discusses DC biasing, frequency response, and feedback. Chapters 4 and 5 cover differential amplifiers and operational amplifier applications, respectively. The final chapter presents oscillator circuits in detail, from classic feedback-based configurations to advanced designs using negative resistance as well as coupled oscillators. In addition, multivibrators are also covered in this chapter.

Throughout the book, theoretical discussions are supported by circuit diagrams, simulation results, mathematical derivations, and practical examples. While maintaining mathematical rigor, the text emphasizes conceptual clarity and hands-on problem-solving. Each chapter is designed to support both independent learning and classroom instruction, closely aligned with a typical undergraduate electrical engineering curriculum. In addition to serving as a textbook, this work can be used as a supplemental study guide for anyone seeking a clearer understanding of microelectronics.

I would like to extend my gratitude to those who contributed to the development and publication of this second edition. First, I thank the editors at IOP Publishing for inviting me to write this expanded version. I am also grateful to the manuscript reviewers—Brad Deken, Rhianna Johnson, Madison Smith, and Collin Stinson—for their constructive comments and suggestions. Special thanks go to my wife, Qin Zhong, for her ongoing support and encouragement throughout the writing process. Finally, I wish to thank all my former students who have taken my courses over the past twenty-five years, including *Circuit Analysis*, *Semiconductor Devices*, *Electromagnetics*, *Electronic Circuits*, *Analog Integrated Circuit Design*, *Digital System Design*, *RF/Microwave Circuit Design*, *Signals and Systems*, and *Computer Systems and Assembly Language*. Preparing lectures and engaging with students has continually deepened my own understanding and inspired me to refine my teaching methods. My experience affirms that most students can master electronic circuits when teaching and learning both follow the right approach.

This book is motivated by a desire to help more students overcome the challenges of learning electronic circuits and to inspire future success in the field of electronics. I hope it serves as a valuable companion on your academic and professional journey into this fascinating and ever-evolving field.

Yumin Zhang
Southeast Missouri State University
Cape Girardeau, MO 63701, USA
July 22, 2025

Author biography

Yumin Zhang

Yumin Zhang is a professor in the Department of Engineering and Technology at Southeast Missouri State University. In 1989, he earned an MS degree in physics from Zhejiang University in China and subsequently joined the technical staff at the Institute of Semiconductors, Chinese Academy of Sciences. He received his PhD in electrical engineering from the University of Minnesota in 2000. He later held faculty positions at the University of Wisconsin–Platteville and Oklahoma State University, Stillwater, before joining Southeast Missouri State University in 2007. His academic interests span a wide range of topics, including semiconductor devices, analog and digital electronic circuits, computer systems, dynamic systems, thermodynamics, and engineering education.

Chapter 1

Circuit analysis

Passive circuit elements form the foundation of modern electronic systems and are essential in a broad range of applications. Furthermore, they serve as key components in modeling the behavior of transistors. The three basic passive components are resistors, capacitors, and inductors. Among them, resistors exhibit simple behavior governed by Ohm's law. In terms of energy dynamics, resistors dissipate energy, whereas capacitors and inductors temporarily store and release energy. The current–voltage relationships of capacitors and inductors are more complex, involving differential equations. However, when transformed into the frequency domain, these relationships become linear and take on forms similar to Ohm's law, simplifying analysis.

Interesting and functional circuits—such as filters and resonators—can be constructed using various combinations of these passive elements. In amplifier circuits, the low-frequency response is often shaped by coupling capacitors, while the high-frequency response is influenced by internal capacitances within transistors. In radio-frequency (RF) and microwave applications, inductors play a crucial role, and even the parasitic inductance of transistor packaging must be accounted for.

This chapter lays the foundation for understanding a wide range of electronic circuits by focusing on key components and their dynamic behavior. Section 1.1 explores capacitors and inductors—essential energy storage elements that introduce time dependence into circuits. Section 1.2 introduces RLC circuits, where resistors, inductors, and capacitors interact to produce resonant behavior. To analyze these circuits effectively, the concepts of impedance and admittance are presented in section 1.3, providing a unified framework for handling sinusoidal signals. The next several sections apply these principles to practical circuit configurations: first-order LPFs and HPFs are discussed in section 1.4, followed by the frequency responses of RCR (resistor–capacitor–resistor) modules in section 1.5. Building upon this, second-order filter designs are introduced in section 1.6, enabling sharper transitions between the pass band and the stop band. Sections 1.7 and 1.8 extend the discussion

doi:10.1088/978-0-7503-5512-4ch1

to band-pass filters (BPFs) and band-stop filters (BSFs) of both the first and higher orders, which are crucial in applications such as signal conditioning, noise suppression, and communication systems. Throughout this chapter, emphasis is placed on frequency-domain analysis, circuit behavior interpretation, and design strategies.

1.1 Capacitors and inductors

The symbol for a capacitor is derived from the parallel-plate capacitor model, as shown in figure 1.1. However, the concept of capacitance is much broader: any two conductors separated by an insulating medium form a capacitor. For instance, even the human body can function as a conductor; two people standing close together, without touching, essentially form a capacitor.

The capacitance of a parallel-plate capacitor is given by:

$$C = \varepsilon_r \varepsilon_0 \frac{A}{d}. \tag{1.1}$$

A—area of the plates
d—distance between the plates
ε_r—relative permittivity of the dielectric material between the plates
ε_0—permittivity or dielectric constant of free space (vacuum)

This formula also helps explain the behavior of capacitors in combination. For example, if two capacitors are connected in parallel, the parameter d can be assumed to be the same. The equivalent area is $A = A_1 + A_2$, and thus the total capacitance is $C = C_1 + C_2$. In addition, this equation can be extended to the situation where more than two capacitors are connected in parallel: $C = \sum_{i=1}^{n} C_i$. On the other hand, if n capacitors are in series, it can be assumed that the area A of the plates is the same, and the equivalent distance between the top and bottom plates becomes $d = \sum_{i=1}^{n} d_i$; therefore, the total capacitance can be found using the formula $\frac{1}{C} = \sum_{i=1}^{n} \frac{1}{C_i}$.

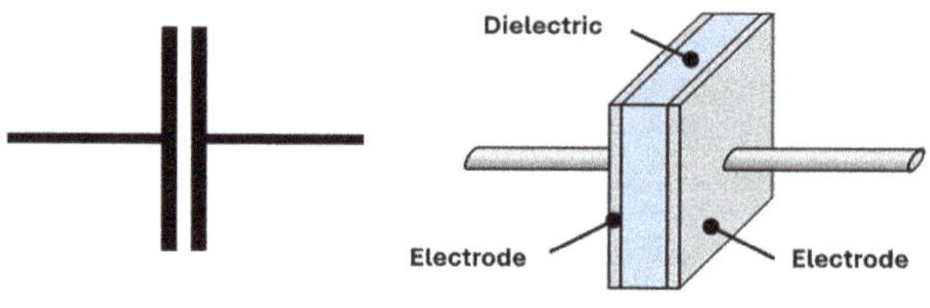

Figure 1.1. Capacitor symbol (left) and a parallel-plate capacitor (right). This [Plate Capacitor DE] image has been obtained by the author from the Wikimedia website where it was made available by [Cepheiden] under a CC BY-SA 3.0 licence. It is included within this book on that basis. It is attributed to [Cepheiden].

Compared with the formulae for resistors in series and parallel, these formulae are just the opposite.

Although parallel-plate capacitors are conceptually simple, they are not suitable for high-capacitance applications. To increase capacitance, flexible conductive materials are rolled into cylindrical shapes, maximizing surface area. Additionally, electrolyte-based capacitors use liquids with higher relative permittivities than those of solids. As shown in figure 1.2, blue and black cylindrical capacitors are electrolytic types with much higher capacitance than ceramic disk capacitors. However, these electrolytic capacitors exhibit polarity and may fail—sometimes violently—if connected in reverse.

To understand capacitor behavior in circuits, a helpful analogy is the hydraulic model, as shown in figure 1.3. Here, the current flowing into a capacitor is analogous to a water flow, and the capacitor acts as a container. The base area of the container corresponds to the *capacitance*, the water level represents the *voltage*, and the volume of the liquid corresponds to the stored *electric charge*. In this hydraulic model, the volume of water can be determined from the other two parameters:

Figure 1.2. Various types of capacitors. Created with GPT-4.0, OpenAI.

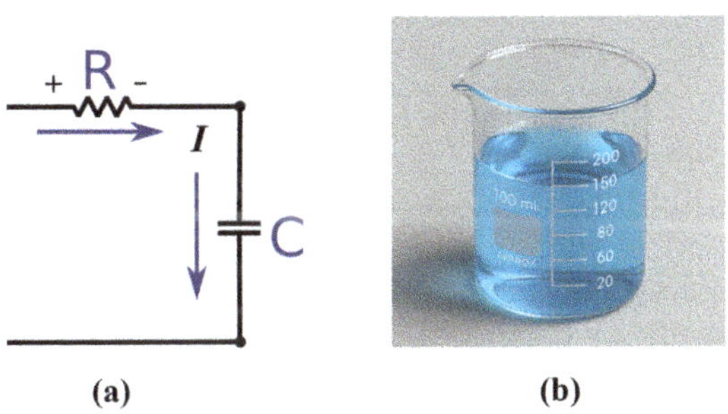

Figure 1.3. Models of capacitors: (a) a capacitor in a circuit, (b) a hydraulic capacitor model. Created with GPT-4.0, OpenAI.

$vol = A \cdot h$. Therefore, the stored electric charge in a capacitor is the product of the capacitance and the voltage:

$$Q = C \cdot V. \tag{1.2}$$

This analogy can be extended further for the energy stored in a capacitor. In this hydraulic model, the potential energy of water is $E_V = mgh_{cm} = (\rho \cdot Vol)g(h/2) = \frac{1}{2}\rho g A h^2$, where $h_{cm} = h/2$ gives the height at the center of mass. If $\rho g = 1$, this expression can be simplified to $E_V = \frac{1}{2}Ah^2$. Likewise, the energy stored in a capacitor can be expressed in a similar manner:

$$E_C = \frac{1}{2}CV^2. \tag{1.3}$$

According to electromagnetic theory, the energy in a capacitor is stored in space, and the energy density can be expressed as $w_e = \frac{1}{2}\varepsilon_r \varepsilon_0 E^2$. For a parallel-plate capacitor, we can assume that the energy is stored between the two plates, and the volume is $vol = Ad$. The electric field there is related to the voltage between the plates: $E = V/d$. With these expressions, we can derive the energy stored in a capacitor: $E_C = w_e \cdot Vol = \frac{1}{2}CV^2$.

Drawing from the hydraulic analogy, we can derive a fundamental relationship between current and voltage in a capacitor. In this model, the volume of water collected in a container corresponds to the electric charge stored in a capacitor, while the flow rate of water represents the electric current. Calculus offers an elegant framework to describe this cumulative behavior: just as the total volume of water is the time integral of the flow rate, the total electric charge $Q(t)$ is the integral of the current $i(t)$, as expressed in equation (1.4). In addition, by substituting equation (1.2) into this equation, we obtain a direct relationship between voltage and current in a capacitor:

$$\begin{aligned} Q(t) &= Q(0) + \int_0^t i_C(\tau)d\tau \\ v_C(t) &= v_C(0) + \frac{1}{C}\int_0^t i_C(\tau)d\tau \end{aligned} \tag{1.4}$$

The first term on the right-hand side represents the initial quantity, while the second term represents the additional quantity added during the given period of time. Taking the derivative of both sides eliminates the constant term, resulting in a more concise equation in a differential format:

$$i_C(t) = C\,\frac{d\,v_C(t)}{dt} \tag{1.5}$$

Electric current is commonly understood as the flow of electric charges. However, it is important to note that no actual electric charge jumps across the gap between the plates of a capacitor. This raises the question of how current is defined in this case. To address this puzzle, James Clerk Maxwell invented the concept of 'displacement current,' which is a pseudo current that flows through a capacitor. The displacement current density is defined as $j_d = \varepsilon_r \varepsilon_0 \frac{dE}{dt}$. In a parallel-plate capacitor, where $E = V/d$ and $i_C = Aj_d$, it can be shown that $i_C = C\frac{dv_C}{dt}$.

In the earlier discussion, capacitors were shown to relate closely to electric fields and thus voltage. Inductors, on the other hand, are more associated with magnetic fields and current. Moreover, using the analogy of a mechanical system, the energy stored in a capacitor is associated with potential energy, while the energy stored in an inductor is equivalent to kinetic energy.

The inductor symbol comes from the solenoid, as shown in figure 1.4. However, in modern applications, inductors include any wire geometry capable of producing magnetic flux. Even a short bonding wire in an RF integrated circuit (IC) layout behaves as a small inductor. Unlike capacitors, inductors always have parasitic resistance due to the resistivity of conductors. For a straight wire, resistance increases linearly with inductance; for coiled solenoids, resistance increases approximately with the square root of inductance. For an ideal solenoid, the inductance is related to the geometric parameters $L = \mu_r \mu_0 AN^2/l$, where N is the number of turns, A is the cross-sectional area, l is the length of the solenoid, and μ_r is the relative permeability of the core material. The factor of N^2 indicates that the inductance is proportional to the square of the wire length.

When compared with capacitors, the mechanical model of inductors may seem less intuitive. In figure 1.5, a water mill is used to illustrate the behavior of an inductor: the current passing through the inductor resembles the flow of water, which is related to the angular velocity of the wheel, and the difference in water levels before and after the water mill corresponds to the voltage across the inductor. The moment of inertia of the water mill is related to the inductance. When the water mill rotates, its kinetic energy is given by $E_K = \frac{1}{2}I_M \cdot \omega^2$, where I_M stands for the moment of inertia of the water mill, which is equivalent to the inductance. By

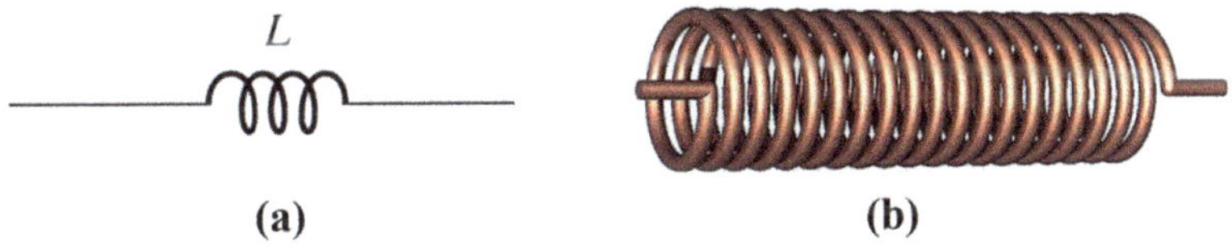

Figure 1.4. (a) Inductor symbol, (b) a solenoid. The [Solenoid-coreless] image (b) has been obtained by the author from the Wikimedia website where it was made available by [Svjo] under a CC BY-SA 3.0 licence. It is included within this book on that basis. It is attributed to [Svjo].

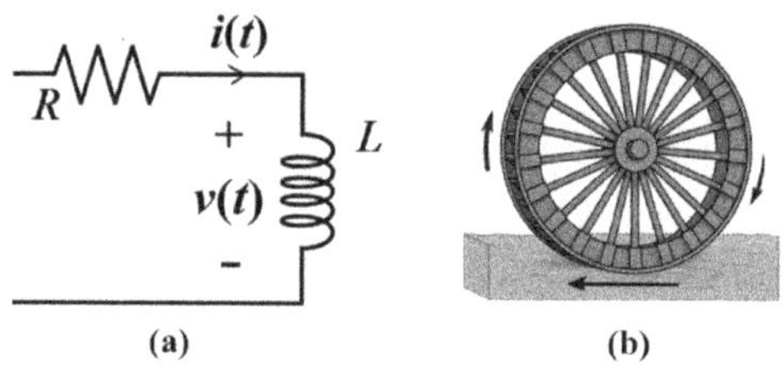

Figure 1.5. Models of inductors: (a) an inductor in a circuit, (b) a mechanical model. Created with GPT-4.0, OpenAI.

converting the mechanical parameters into their electrical counterparts, we can determine the energy stored in an inductor:

$$E_{\mathrm{L}} = \frac{1}{2}LI^2. \tag{1.6}$$

> The energy stored in an ideal solenoid can easily be derived from the magnetic field energy in space. First, the energy density can be expressed as $w_m = \frac{1}{2}\mu_r\mu_0 H^2$, and the magnetic field is a linear function of the current: $H = NI/l$. The volume inside the solenoid is $vol = Al$. Using these expressions, we can derive the energy stored in an ideal solenoid: $E_{\mathrm{L}} = w_m \cdot vol = \frac{1}{2}LI^2$.

The relationship between current and voltage in an inductor can be obtained from the energy equation. Taking a derivative with respect to time, the left-hand side represents the input power: $p_{\mathrm{L}}(t) = i_{\mathrm{L}}(t) \cdot v_{\mathrm{L}}(t)$. By canceling the current on both sides, the following equation can be derived:

$$v_{\mathrm{L}}(t) = L\frac{d}{dt}i_{\mathrm{L}}(t) \tag{1.7}$$

Indeed, the relationship between current and voltage in a capacitor can be obtained in a similar manner. However, the derivation using the hydraulic model provides a more intuitive understanding. While magnetic charge does not exist, an equivalent counterpart to the amount of electric charge can be defined, known as magnetic flux. Most inductors have multiple turns; in this case, a new parameter is used, known as *magnetic flux linkage*: $\Lambda = N\Phi = L \cdot I$, where $L = NL_0$. If each turn is equivalent to a small battery, the total voltage across the inductor is just like N small batteries in series, resulting in the terminology of 'linkage.'

> In classical physics, magnetic flux is considered a continuous quantity. In contrast, electric charge is discrete because electrons cannot be divided. However, in quantum mechanical systems such as superconductors, magnetic flux becomes quantized. This means it is possible to count the number of magnetic flux quanta, which can be expressed in terms of fundamental physical constants: $\Phi_0 = h/2e$. This leads to technologies such as superconducting quantum interference devices (SQUIDs) that can detect minute magnetic changes.

1.2 RLC circuits

When a capacitor and an inductor are connected together, as shown in figure 1.6(a), they form an LC resonant circuit. As explained in earlier sections, capacitors store energy in electric fields, which is analogous to potential energy, while inductors store energy in magnetic fields, which is analogous to kinetic energy. When these two elements are combined, the resulting system behaves like a mass–spring oscillator. If initial energy is supplied—either by applying a voltage across the capacitor or a current through the inductor—the system enters a state of free oscillation, exchanging energy back and forth between the capacitor and the inductor. Figure 1.6(b) shows the waveforms of the voltage across the capacitor and the current flowing into the capacitor, which have a 90° phase shift between them.

As we know, the average power consumed by a device is: $P = I_{\text{rms}} V_{\text{rms}} \cos(\theta)$, where the *rms* values of current and voltage are used. With a phase shift of 90° between current and voltage, the power consumption becomes zero, which is required for capacitors as well as inductors.

In a mass–spring system, the oscillation frequency is determined by the equation $\omega_0 = \sqrt{k/m}$, where k stands for the spring constant. Following the mass–spring analogy, the resonant frequency in an LC circuit can be calculated using an equation with a similar format:

$$\omega_0 = \frac{1}{\sqrt{LC}} \Leftrightarrow f_0 = \frac{1}{2\pi\sqrt{LC}}. \tag{1.8}$$

In the analogy of a mass–spring system, inductance corresponds to mass, while capacitance is related to the reciprocal of the spring constant. Compared with the formula for angular frequency, the formula for the resonant frequency is less concise. However, its relationship with the period is straightforward: $T = 1/f$.

In real-world applications, inductors are never ideal—they always include parasitic resistance. In figure 1.7(a), the series resistor R_1 stands for this resistance. As we know, resistance causes energy loss, which converts electrical energy into heat. As shown in figure 1.7(b), the amplitude of the capacitor's voltage decays over time, forming a damped sine wave, which can be expressed as:

$$v_c(t) = V_0 e^{-\alpha t} \cos(\omega_d t + \phi). \tag{1.9}$$

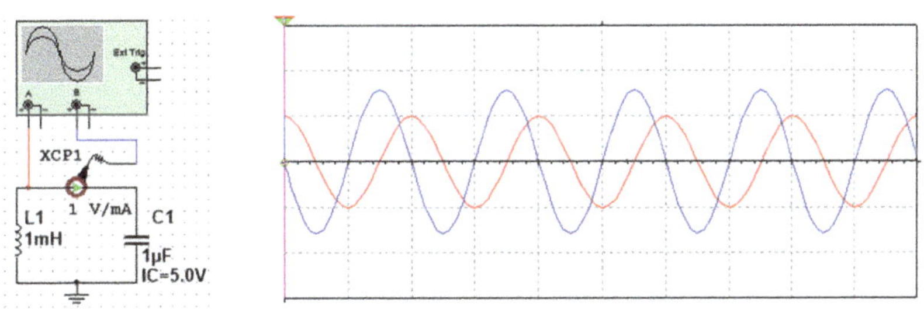

(a) (b)

Figure 1.6. (a) LC resonant circuit, (b) voltage and current waveforms.

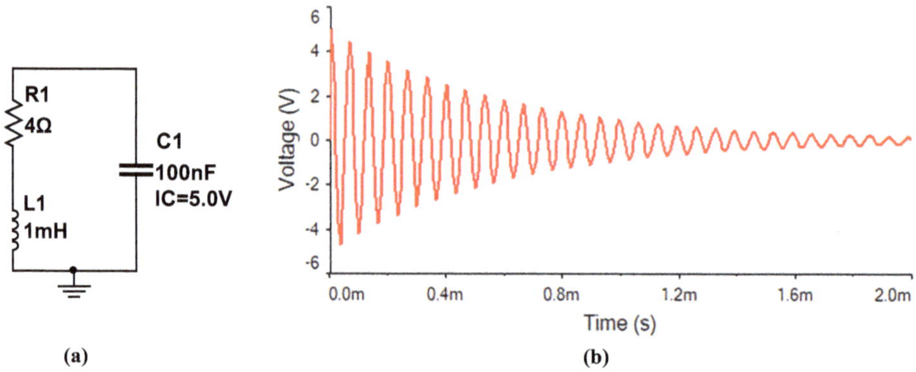

Figure 1.7. LC circuit with parasitic resistance: (a) circuit, (b) waveform.

The parameters in this expression can be found by solving a differential equation.

- $\alpha = R/(2L)$—damping coefficient
- $\omega_\mathrm{d} = \sqrt{\omega_\mathrm{o}^2 - \alpha^2}$—damped angular frequency
- ϕ—phase angle

In the damped oscillator circuit, the frequency ω_d is slightly lower than the resonant frequency ω_o. However, these two parameters are very close for low-loss systems, so they are often not distinguished: $\omega_\mathrm{d} \approx \omega_\mathrm{o}$. In the circuit shown in figure 1.7(a), $\alpha = 2 \times 10^3$, which is much less than the first term, $\omega_\mathrm{o} = 10^5$. However, if a large capacitor is used, ω_o becomes lower, and the shift in frequency can be significant.

Although the frequency shift can often be ignored, the role of the damping coefficient, which is proportional to the parasitic resistance, is very important. Therefore, the greater the resistance, the faster the dissipation of energy, since the power consumption is proportional to the resistance: $p(t) = R \cdot i^2(t)$.

To quantify the damping behavior, a parameter known as the *quality factor* (Q) is introduced, which is proportional to the ratio of energy stored to the energy dissipated per cycle, and it is defined as:

$$Q = \omega_0 \frac{E_\mathrm{st}}{P_\mathrm{ds}} \Rightarrow Q = \omega_0 \tau_0 = 2\pi \frac{\tau_0}{T}. \tag{1.10}$$

In this equation, E_st represents the energy stored in the system and P_ds stands for the power of dissipation. The ratio of these two quantities determines the decay rate of the waveform, as shown in figure 1.7(b). Assume that the initial current going through the inductor is I_o; then, the energy stored in the inductor is $E_\mathrm{st} = 0.5LI_o^2$. On the other hand, the average dissipated power is $P_\mathrm{ds} = 0.5RI_0^2$ for a sine wave. Therefore, the ratio of these two quantities gives us the time constant $\tau_0 = E_\mathrm{st}/P_\mathrm{ds} = L/R$, which is consistent with the definition of the time constant in an *RL* transient circuit. In addition, the quality factor can be expressed with these three circuit elements:

$$Q = \omega_0 \tau_0 = \sqrt{L/C}/R.$$

In spectral analysis, the quality factor (Q) indicates how sharply a system responds near its resonant frequency. It is defined as $Q = f_o/\Delta f$, where f_o is the resonant frequency and Δf is the bandwidth (BW). A higher Q means a narrower, sharper peak in the frequency response. This reflects greater frequency selectivity and lower energy loss, making high-Q systems ideal for oscillators and precise tuning applications. Conversely, low-Q systems exhibit broader peaks, indicating less selectivity. Thus, the sharpness of the resonance peak directly visualizes the system's Q factor.

1.3 Impedance and admittance

Most desktop computers in the lab are connected to the external world using two main cables: one for the power supply and the other for data transmission (e.g. Ethernet). Interestingly, both power and information are transmitted through sine waves, which are particularly important among all types of AC signals due to their unique properties in physical systems. Figure 1.8(a) illustrates a general potential energy profile, where the position L represents a stable equilibrium point. Near this location, the potential energy function can be approximated using a Taylor series expansion, where the origin of the x-axis is set at the equilibrium position:

$$y(x) = y(0) + y' \cdot x + \frac{1}{2!}y'' \cdot x^2 + \frac{1}{3!}y''' \cdot x^3 + \cdots \approx \frac{1}{2}kx^2. \tag{1.11}$$

The first term, being a constant, is irrelevant and can be set to zero. The second term disappears, as the first derivative at equilibrium positions is zero. Neglecting the higher-order terms, we are left with the quadratic term, as illustrated in figure 1.8(b).

This potential energy profile shown in figure 1.8(b) is commonly employed to describe the well-known mass–spring system, where the potential energy is a quadratic function of displacement, and the solution of this system is a sinusoidal function:

$$U(x) = \frac{1}{2}k\, x^2 \Rightarrow x(t) = A_0 \cos(\omega t + \theta_0), \tag{1.12}$$

where the parameters A_0 and θ_0 can be determined by the initial conditions.

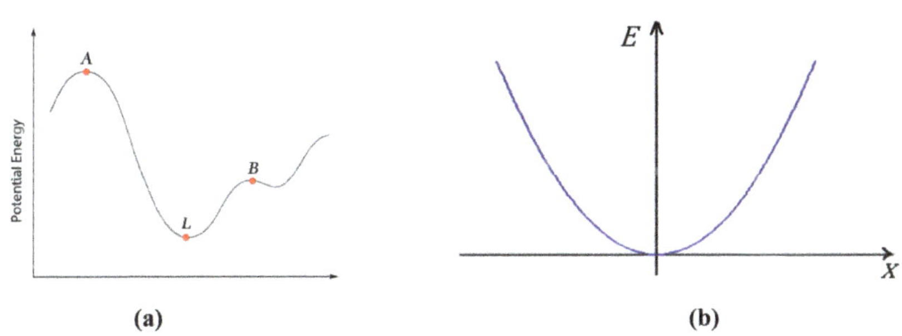

Figure 1.8. Potential-energy diagrams: (a) equilibrium positions, (b) parabolic approximation. Created with GPT-4.0, OpenAI.

This example reveals that sine and cosine functions naturally emerge in the analysis of systems near equilibrium. A familiar real-world example is the vibration of a plucked guitar string, which produces sinusoidal oscillations. In fact, many natural systems—ranging from atomic vibrations to tectonic motion—oscillate near equilibrium positions, often resulting in sinusoidal behavior.

Although waveforms such as square and triangular waves are commonly used, they tend to distort during long-distance transmission. According to Fourier analysis, any periodic waveform can be represented as a sum of sine and cosine components. However, most transmission media exhibit dispersion, meaning that the wave velocity varies with frequency. As a result, different frequency components travel at different speeds, causing phase shifts and distortion. This is a key reason why sinusoidal signals are preferred in communication systems—they consist of a single frequency component and can preserve their shape over long distances.

> Dispersion in physics refers to the phenomenon where the velocity of a wave depends on its frequency. This terminology originated from the separation of white light into colors through a prism. However, it is not limited to optics; instead, it occurs in various wave types, including light, sound, and water waves. Dispersion plays a critical role in optics, acoustics, and communication systems, where it affects signal clarity and timing by altering the shape of waveforms during transmission.

1.3.1 Exponential function

While sinusoidal functions are fundamental, they are inconvenient for solving differential equations directly. For example, the derivative of a sine function is a cosine function, not a multiple of itself. To simplify analysis, sine and cosine waves are often represented using complex exponentials via Euler's identity:

$$e^{i\omega t} = \cos(\omega t) + i\,\sin(\omega t). \tag{1.13}$$

To illustrate this visually, imagine a rotating turntable spinning counterclockwise at an angular speed of ω, as illustrated in figure 1.9. A can placed at the edge casts

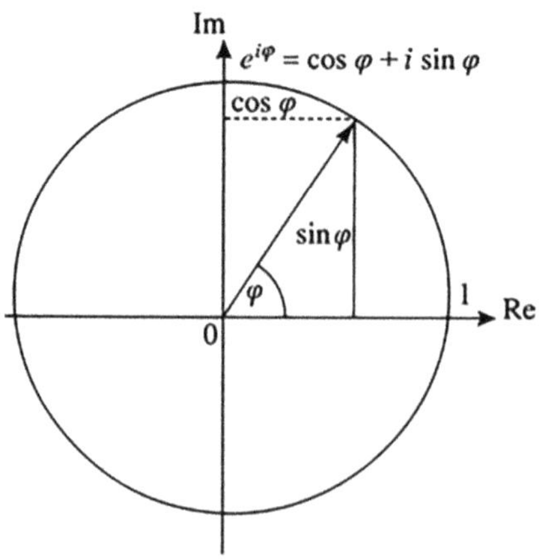

Figure 1.9. Exponential function vs sine and cosine functions. Created with GPT-4.0, OpenAI.

shadows on two perpendicular walls, resulting in sinusoidal projections—cosine and sine waves, respectively.

Because exponential functions are eigenfunctions of the differential operator, their derivatives remain proportional to the original function. This greatly simplifies circuit analysis. In electrical engineering, the imaginary unit ***i*** is replaced by ***j*** to avoid confusion with electric current. For consistency, the cosine function—the real component of the exponential function—is used for the following conversion:

$$e^{j(\omega t+\theta_0)} \leftrightarrow \cos(\omega t + \theta_0). \tag{1.14}$$

In this way, electrical signals to be expressed compactly using phasors and exponentials, and the concepts of impedance and admittance can be introduced.

1.3.2 Impedance and admittance

By representing voltages and currents as exponential functions, we can derive simple linear relationships for capacitors and inductors—much like Ohm's law for resistors. For a capacitor:

$$v(t) = V_0 e^{j\omega t} \Rightarrow i(t) = C\frac{dv(t)}{dt} = (j\omega C)v(t). \tag{1.15}$$

This suggests that the capacitor behaves like a *complex resistor* with the following expressions for impedance and admittance:

$$Z_C = \frac{1}{j\omega C}, \quad Y_C = j\omega C \tag{1.16}$$

Here, Z_C is the *impedance* and Y_C is the *admittance*. Impedance and admittance are generalized concepts that extend resistance and conductance to AC analysis. In general, they are complex numbers, with real and imaginary components:

$$\begin{aligned} Z &= R + jX \\ Y &= G + jB, \end{aligned} \tag{1.17}$$

where:

- R is resistance
- X is reactance
- G is conductance
- B is susceptance.

Similarly, a linear relationship between current and voltage can also be obtained for an inductor:

$$v(t) = (j\omega L)i(t) \Rightarrow Z_L = j\omega L, \quad Y_L = \frac{1}{j\omega L}. \tag{1.18}$$

These impedance and admittance values are summarized in table 1.1.

Table 1.1. Impedance and admittance of capacitors and inductors.

Device	Impedance (Ω)	Admittance (S)
Capacitor	$Z_C = \frac{1}{j\omega C}$	$Y_C = j\omega C$
Inductor	$Z_L = j\omega L$	$Y_L = \frac{1}{j\omega L}$

1.3.3 Asymptotic behavior at extremes of frequency

Analyzing the behaviors of capacitors and inductors at extremely low and high frequencies proves valuable in comprehending various passive filters.

- At the low-frequency limit ($\omega \rightarrow 0$):
  - $Z_C \rightarrow \infty$: a capacitor acts like an **open circuit**
  - $Z_L \rightarrow 0$: an inductor behaves like a **short circuit**
- At the high-frequency limit ($\omega \rightarrow \infty$):
  1. – $Z_C \rightarrow 0$: a capacitor behaves like a **short circuit**
  2. – $Z_L \rightarrow \infty$: an inductor acts like an **open circuit**

These asymptotic behaviors are summarized in table 1.2.

In DC circuits, capacitors can generally be treated as open circuits, so they are used in discrete amplifier circuits to separate DC and AC signals. Essentially, capacitors allow AC signals to pass through while blocking DC currents—this is called *AC coupling*. Inductors, on the other hand, behave like short circuits for DC but have significant impedances at high frequencies, which allows them to be employed as *RF chokes* in amplifier circuits.

> As discussed in the first section, capacitors and inductors are related to electric and magnetic fields, respectively. While resistors are also related to electric fields, one may wonder whether there is a counterpart for them. In 1971, Dr Leon Chua proposed a new passive device to fill this vacancy—the memristor. It was not until more than three decades later, in 2007, that the first solid-state memristor was demonstrated by a research team from a lab at Hewlett-Packard Company. Memristors hold great potential for applications in neural networks and artificial intelligence.

Table 1.2. Asymptotic frequency responses of large capacitors and inductors.

Device	Low frequency	High frequency
Capacitor	Open circuit	Short circuit
Inductor	Short circuit	Open circuit

1.4 First-order low-pass and high-pass filters

Electronic filters are fundamental components designed to selectively allow or block certain frequencies of electrical signals. Hence, they play a crucial role in various applications, ranging from audio systems and telecommunications to signal processing and instrumentation. Due to their ability to enhance signal quality, remove noise, separate different frequency components, and extract specific information, electronic filters are essential tools for achieving the desired performance in numerous electronic systems.

Two of the most basic and widely used filters are the low-pass filter (LPF) and the high-pass filter (HPF). These can be easily implemented using resistors combined with either capacitors or inductors. Their frequency responses are illustrated in figure 1.10. LPFs allow low-frequency signals to pass through without significant attenuation while blocking high-frequency signals. HPFs work in just the opposite way: the pass band is in the high-frequency domain, while the stop band is in the low-frequency domain. Ideally, there would be an abrupt transition in the transfer function between the pass band and the stop band. However, a transition region exists for all real filters. Generally, higher-order filters exhibit steeper slopes, with the trade-off being more complicated circuits. Achieving an optimal balance between filter order, transition characteristics, and circuit complexity is crucial in filter design.

1.4.1 RC low-pass filter

In AC circuits, all three passive circuit elements can be treated in the same way. For example, the RC LPF depicted in figure 1.11(a) can be converted into the form of figure 1.11(b), which resembles a voltage divider circuit with two resistors. The transfer function is defined as the ratio of the output signal to the input signal, which can be readily derived from the equivalent circuit in figure 1.11(b).

$$H(j\omega) = \frac{\tilde{V}_{\text{out}}}{\tilde{V}_{\text{in}}} = \frac{Z_{\text{C}}}{R + Z_{\text{C}}} = \frac{1}{1 + j\omega RC} = \frac{1}{1 + j\omega/\omega_{\text{c}}} \tag{1.19}$$

In this equation, the cutoff angular frequency is defined as: $\omega_{\text{c}} = 1/(RC)$. It is well known that the product of RC is a time constant, so its reciprocal corresponds to a frequency.

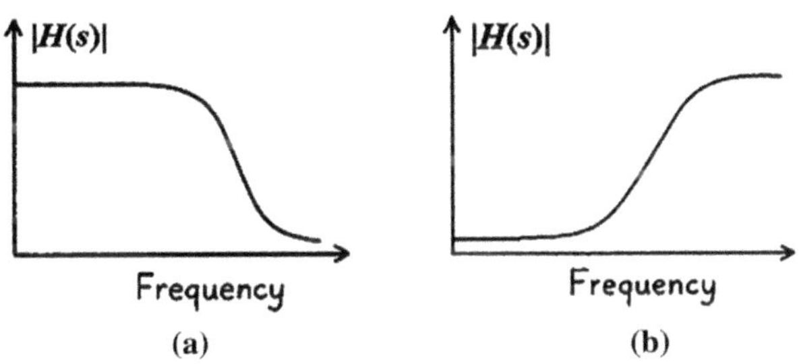

Figure 1.10. Transfer characteristics: (a) low-pass filter (LPF), (b) high-pass filter (HPF). Created with GPT-4.0, OpenAI.

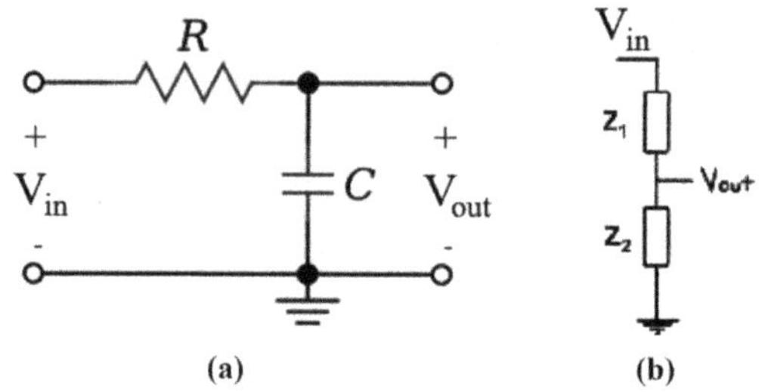

Figure 1.11. (a) RC LPF circuit, (b) equivalent circuit with impedance. Created with GPT-4.0, OpenAI.

1.4.2 Bode plot and frequency response

As we know, a complex number can be expressed in two different ways: Cartesian and polar formats. Similarly, a complex function can also be converted into two real functions in these two approaches. For transfer functions, the polar format is more popular.

$$H(j\omega) = |H(j\omega)|e^{j\theta(\omega)}. \tag{1.20}$$

For the RC LPF, these two components of the transfer function are:

$$|H(j\omega)| = \frac{1}{\sqrt{1 + (\omega/\omega_c)^2}}, \quad \angle\,\theta = -\tan^{-1}(\omega/\omega_c). \tag{1.21}$$

This transfer function can be simplified with approximations in different frequency domains:

- $\omega = \omega_c$: $H(\omega) = 1$
- $\omega = \omega_c$: $H(\omega) = \frac{1}{1+j} = \frac{1}{\sqrt{2}}\ \angle - 45^\circ$
- $\omega = \omega_c$: $H(s) \approx -j\omega_c/\omega = (\omega_c/\omega)\ \angle - 90^\circ$.

Plots of the magnitude and phase functions are called *Bode plots*, and this approach is an important way to analyze the frequency response of filters and amplifiers. Bode plots can be conveniently obtained using the circuit simulation software package NI Multisim, which includes a virtual instrument called the *Bode Plotter*, shown at the top of figure 1.12(a). Bode plots of a simple RC circuit are depicted in figure 1.12(b), with the cutoff frequency indicated by the cursor, estimated to be approximately 10 kHz. The top and bottom plots depict the magnitude and the phase, respectively. The top plot in figure 1.12(b) shows the magnitude of the transfer function in decibels (dB), a widely used unit for describing numbers across a broad range.

> Hendrik Wade Bode (1905–82) was a notable American engineer, researcher, inventor, and scientist who made significant contributions to the fields of modern control theory and electronic telecommunications. His pioneering work revolutionized the content and methodology of these disciplines. During his tenure at Bell Labs in the 1930s, Bode devised a straightforward yet precise technique for plotting gain and phase-shift characteristics, which became a valuable tool in circuit theory and control theory.

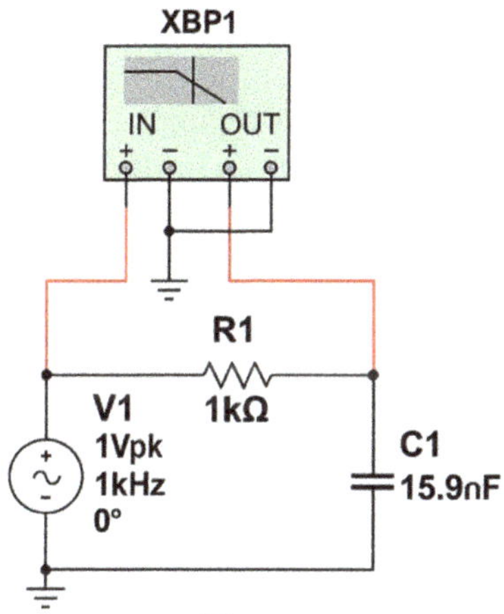

(a)

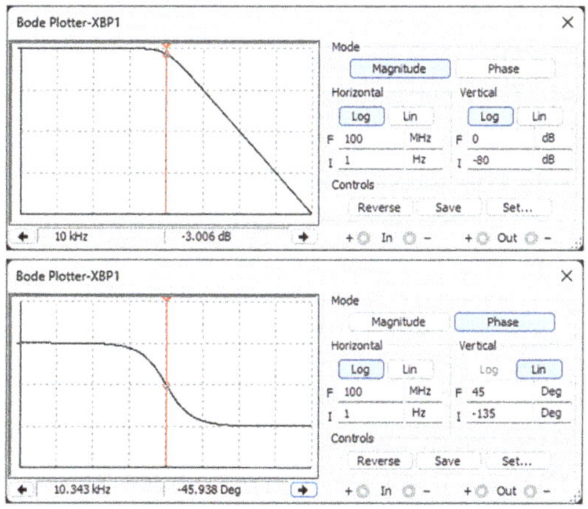

(b)

Figure 1.12. (a) RC LPF circuit, (b) Bode plots.

Table 1.3. Decibel conversion table.

$\|H(\omega)\|$	0.01	0.1	$\frac{1}{2}$	$\frac{1}{\sqrt{2}}$	1	$\sqrt{2}$	2	10	100
dB	−40	−20	−6	−3	0	3	6	20	40

The term 'decibel' originated from Bell Labs—the 'bel' was an abbreviation of 'Bell' and the 'deci' means one-tenth. When measuring power, the conversion formula used to obtain the dB value is $10\log_{10}(P/P_o)$, where P_o represents the reference power. For example, the minimum power of sound detectable by human ears can serve as the reference power P_o, and when sound power increases, it can be quantitatively expressed in dB using this formula. If the sound power reaches excessive levels, it can potentially cause damage to human hearing, and this limit is at 120 dB. This number is called a *dynamic range*, which is a crucial parameter of sensors and transducers. In the realm of human perception, our eyes possess an astonishing dynamic range of 140 dB, surpassing the capabilities of typical cell phone cameras by a considerable margin. Consequently, the limitations of our cell phones prevent them from adequately capturing the breathtaking beauty we witness during the enchanting moments of dawn and dusk.

In electric circuits, voltage and current are parameters that can be measured easily. However, it is important to note that the formula needs to be changed due to $P = I^2R = V^2/R$:

$$\left|H(j\omega)\right|(dB) = 20\log_{10}\left|H(j\omega)\right|. \tag{1.22}$$

For the reader, adapting to this novel way to describe the transfer function may require some time. To assist beginners in navigating this concept, table 1.3 can serve as a valuable resource.

Due to its practical application, a common and helpful approximation is to convert a ratio of two into 6 dB, which is musically termed an *octave*. In a similar vein, 3 dB is an approximation for the ratio of $\sqrt{2}$. In addition, this approximation extends to their reciprocals for −6 and −3 dB. As shown in figure 1.12(b), at the cutoff frequency, the magnitude of the transfer function is −3 dB, which is a common feature of first-order filters.

1.4.3 RC high-pass filter

If the positions of the resistor and the capacitor in the LPF circuit are exchanged, the LPF transforms into an HPF, as demonstrated in figure 1.13. These two circuits can easily be confused by beginners; one can utilize the asymptotic behavior of the capacitor to differentiate between them. For instance, at low frequencies, the capacitor behaves as an open circuit, which blocks the signal in the HPF circuit but does not cause any attenuation to the signal in the LPF circuit.

Using the same approach, the transfer function of this HPF circuit can be derived:

$$H(\omega) = \frac{\widetilde{V}_{\text{out}}}{\widetilde{V}_{\text{in}}} = \frac{R}{R + Z_C} = \frac{j\omega RC}{1 + j\omega RC} = \frac{j\omega/\omega_c}{1 + j\omega/\omega_c}. \tag{1.23}$$

This transfer function can be simplified with approximations in different frequency domains:

- $\omega = \omega_c$: $H(\omega) \approx j\omega/\omega_c = (\omega/\omega_c)\ \angle 90^\circ$
- $\omega = \omega_c$: $H(\omega) = \frac{j}{1+j} = \frac{1}{\sqrt{2}}\ \angle 45^\circ$
- $\omega = \omega_c$: $H(s) = 1$.

Comparing the Bode plots of the LPF and the HPF, the magnitude plots appear as mirrored images, while the phase plots share the same shape, except for a 90° phase shift originating from the j term in the numerator of the HPF transfer function. However, there is one key similarity between these two phase plots: there is no phase shift within the pass band. In addition, beyond the edge of the pass band, the

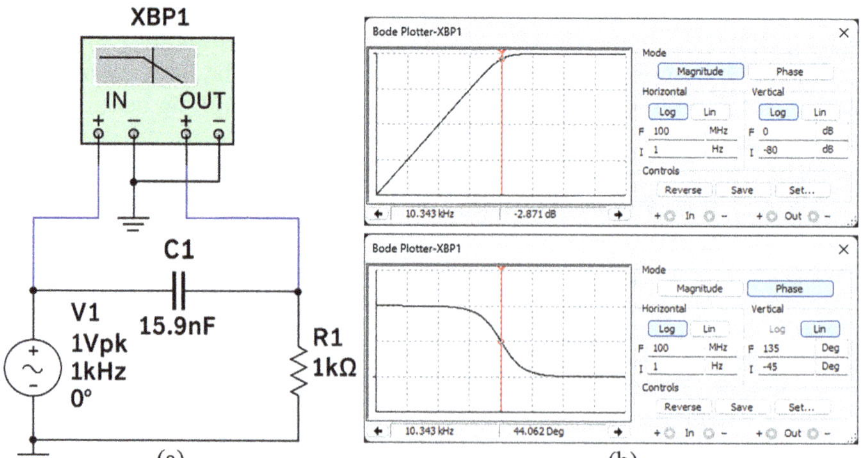

(a) (b)

Figure 1.13. (a) RC HPF circuit, (b) Bode plots.

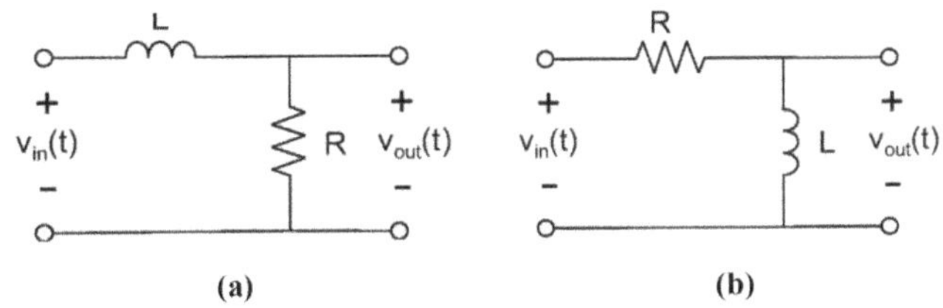

Figure 1.14. First-order RL filters: (a) LPF circuit, (b) HPF circuit.

magnitudes of the transfer functions decrease at a rate of −20 dB/dec all the way down, while the phase only undergoes change in the frequency domain from 0.1 to $10\omega_c$. In simple terms, the phase remains constant outside this domain.

1.4.4 RL low-pass and high-pass filters

If the capacitor is replaced with an inductor, the circuits still function as filters, but their types are reversed, as shown in figure 1.14. The asymptotic behavior of the inductor can also aid in distinguishing between these two filter circuits. At very low frequencies, the inductor behaves similarly to a short circuit, while at very high frequencies, it behaves as an open circuit. Therefore, the circuit in figure 1.14(a) is an LPF, while the circuit in figure 1.14(b) is an HPF. With the cutoff frequency defined as $\omega_c = R/L$, the transfer functions are identical to those of RC filters.

As discussed in the second section, inductors inherently exhibit parasitic resistance, which leads to energy loss and reduced efficiency, particularly in high-frequency applications. Additionally, inductors tend to be physically bulky and are difficult to integrate into modern IC designs due to their size and magnetic field interference. In contrast, capacitors are more compact, cost-effective, and easier to fabricate on semiconductor chips. For these reasons, RC filters are generally preferred over RL filters in practical circuit design, especially in compact and high-density systems such as mobile devices and digital electronics.

1.5 Frequency responses of RCR circuit modules

In this section, we examine four circuit configurations composed of two resistors and one capacitor, each demonstrating distinct and instructive frequency response characteristics. Two of these circuits are relatively straightforward, as they can be transformed into familiar first-order LPFs and HPFs using basic circuit transformations. The other two—commonly known as *lead* and *lag* compensators—are more nuanced and play an important role in amplifier design.

1.5.1 RCR low-pass filter

The circuit at the top of figure 1.15(a) resembles a standard LPF, but a load resistor R_2 is included. To simplify analysis, we can follow a two-step approach:

1. Swap the positions of the capacitor C and resistor R_2, which does not change the overall behavior.
2. Apply Thévenin's theorem to the subcircuit on the left-hand side of the capacitor.

This transformation yields the circuit shown at the bottom of figure 1.15(a). The Thévenin equivalent resistance is: $R12 = R1 + R2$, and the equivalent voltage source becomes $V_2 = V_1[R_2/(R_1 + R_2)]$.

Figure 1.15(b) shows the Bode plots of the original circuit with the output taken from the node above the capacitor. At first glance, these plots resemble those of the classic RC LPF shown in figure 1.12(b). However, a subtle but important distinction exists: the pass-band gain is –6 dB instead of 0 dB. This is because, at low frequencies, the capacitor behaves like an open circuit, and the two resistors form a simple voltage divider, resulting in a transfer function of 0.5 (−6 dB). On the other hand, the phase plot remains identical to that of a standard first-order LPF.

1.5.2 RCR high-pass filter

Similarly, the circuit at the top of figure 1.16(a) resembles an HPF, but with an additional resistor R_1 preceding the capacitor, which could be the internal resistance of a signal source. This circuit can be simplified by swapping R_1 and C and combining the two resistors into a single equivalent resistor: $R12 = R1 + R2$. The resulting configuration behaves as a standard RC HPF in combination with a voltage divider.

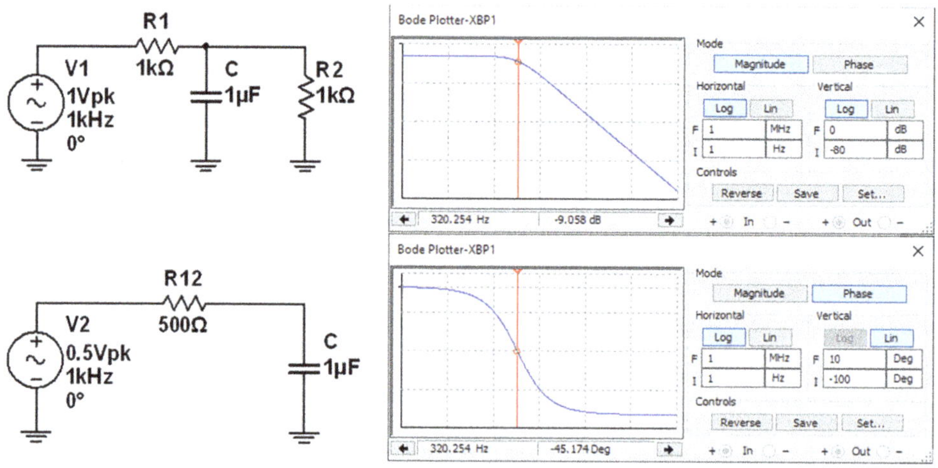

(a) (b)

Figure 1.15. RCR LPF: (a) circuits, (b) Bode plots.

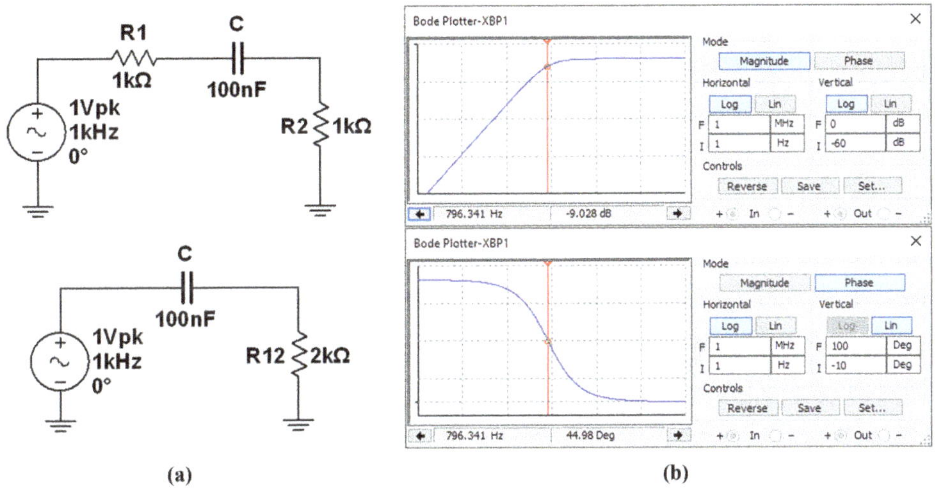

Figure 1.16. RCR HPF: (a) circuits, (b) Bode plots.

The Bode plots shown in figure 1.16(b) correspond to the original circuit presented in the upper section of figure 1.16(a), with the output signal taken at the top of R_2. Compared to a standard RC HPF, like the one depicted in figure 1.13(b), there is a notable difference in the pass-band transfer function, which is at −6 dB. At very high frequencies, the capacitor behaves as a short circuit, effectively turning this circuit into a voltage divider. As a result, the transfer function equals 0.5 (−6 dB) within the pass band.

1.5.3 Pole-zero analysis

To achieve a deeper understanding of filter behaviors, it is often beneficial to place concepts within a broader context. Therefore, the one-dimensional variable $j\omega$ can be extended into a two-dimensional complex variable $s = \sigma + j\omega$. With this approach, the impedances of capacitors and inductors can be expressed as $Z_{\mathrm{C}} = 1/sC$ and $Z_{\mathrm{L}} = sL$, respectively. Therefore, the transfer functions of the two first-order filters can be expressed in new forms:

$$
\begin{aligned}
H_{\mathrm{LPF}}(s) &= \frac{1/sC}{R + 1/sC} = \frac{\omega_c}{s + \omega_c} \Rightarrow H_{\mathrm{LPF}}(s) = \frac{\omega_c}{s - p} \\
H_{\mathrm{HPF}}(s) &= \frac{R}{R + 1/sC} = \frac{s}{s + \omega_c} \Rightarrow H_{\mathrm{HPF}}(s) = \frac{s}{s - p}.
\end{aligned} \tag{1.24}
$$

In these equations, $p = -\omega_c$ is called a *pole*, which refers to the root of an equation where the denominator equals zero. For stable systems, the poles should not be in the right half-plane, $\mathrm{Re}(s) \leqslant 0$, or the output signal will experience an exponential increase over time. Similarly, a *zero* is defined as the root of an equation where the numerator equals zero. Figure 1.17 illustrates a complex plane, displaying the location of the pole alongside the related complex vectors. As the frequency varies,

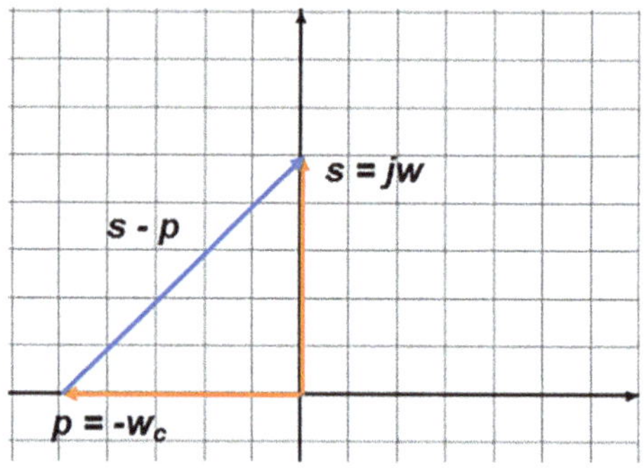

Figure 1.17. Pole location and complex vector in the complex plane.

the trajectory of the variable *s* moves along the imaginary axis, since $s = j\omega$. The denominator of the transfer function, denoted by $s - p$, can be represented by the vector on the complex plane, as demonstrated in figure 1.17.

In the case of the LPF, the numerator remains constant, with a value of ω_c; thus, the characteristics of the transfer function are solely determined by the denominator. The scenario depicted in figure 1.17 corresponds to the condition $\omega = \omega_c$. In this case, $|s - p| = \sqrt{2}\omega_c$ and $\angle(s - p) = 45°$; hence, the transfer function is equal to $(1/\sqrt{2})\ \angle{-45°}$.

– At very low frequencies ($\omega = \omega_c$):

$$|s - p| \approx \omega_c, \ \angle(s - p) \approx 0°, \ H(s) \approx 1.$$

– At high frequencies ($\omega = \omega_c$):

$$|s - p| \approx \omega, \ \angle(s - p) \approx 90°, \ H(s) \approx -j\omega_c/\omega$$

1.5.4 Lead compensator

In addition to the two filter circuits discussed above, there are alternative ways to construct a circuit utilizing two resistors and one capacitor. The circuit illustrated in figure 1.18(a) is referred to as a *lead compensator* circuit, and its frequency response is presented in figure 1.18(b). First, an asymptotic analysis can be conducted: at very low frequencies, the capacitor behaves as an open circuit, causing the circuit to function as a voltage divider with a transfer function of 0.5. Conversely, at very high frequencies, the capacitor acts like a short circuit, resulting in a transfer function of unity. These asymptotic values are confirmed by the simulated Bode plots displayed in figure 1.18(b).

Between these two extreme cases, the behavior of this circuit can be analyzed using the transfer function of a generalized voltage divider, where Z_1 and Y_1 stand for the impedance and admittance of R_1 and C_1 in parallel, respectively:

$$H(s) = \frac{R_2}{Z_1 + R_2} = \frac{Y_1 R_2}{1 + Y_1 R_2} = \frac{s + \omega_1}{s + \omega_2} = \frac{s - z}{s - p}. \tag{1.25}$$

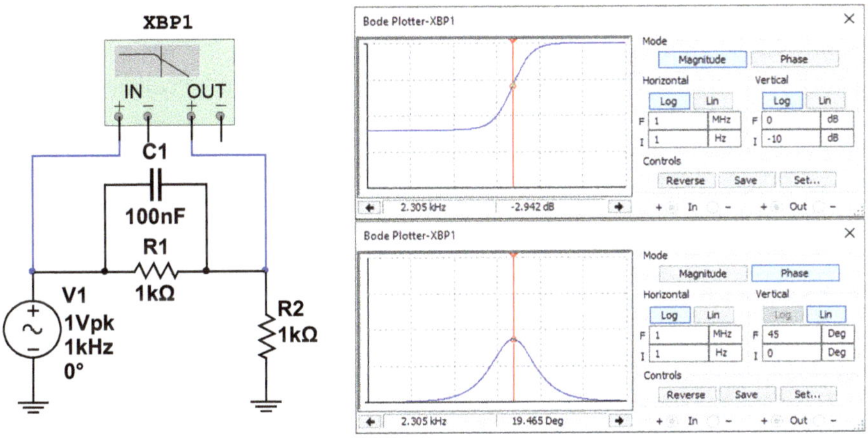

(a) (b)

Figure 1.18. Lead compensator: (a) circuit, (b) Bode plots.

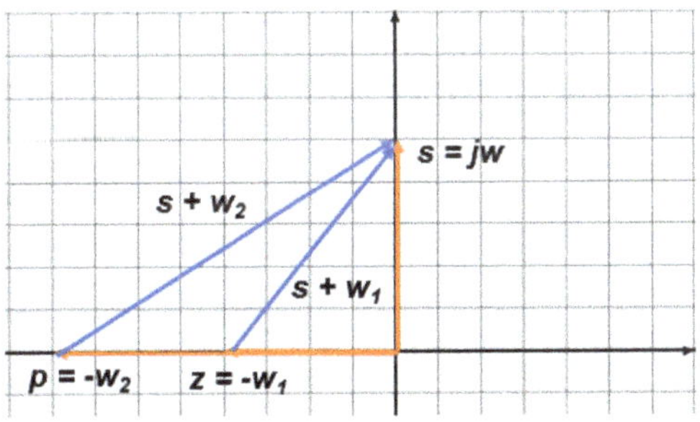

Figure 1.19. Pole-zero diagram of a lead compensator circuit.

In this equation, $\omega_1 = 1/(R_1C_1)$ and $\omega_2 = 1/[(R_1\|R_2)C_1]$. It can be inferred that the first frequency is lower than the second, $\omega_1 < \omega_2$, since $R_1 > R_1\|R_2$. In other words, the pole frequency is higher than the zero frequency, and the pole-zero diagram is depicted in figure 1.19. At very low frequencies, the phase angles of both the numerator and the denominator approach zero, and the magnitude of the transfer function becomes ω_1/ω_2. Conversely, at very high frequencies, the phase angles of both the numerator and the denominator approach 90°, and their magnitudes are essentially the same, so the transfer function tends to approach unity.

One of the primary applications of the lead compensator circuit is to enhance amplifier stability, with a particular focus on the phase shift. Referring to equation (1.25), the phase of the transfer function is $\theta = \theta_1 - \theta_2 = \tan^{-1}(\omega/\omega_1) - \tan^{-1}(\omega/\omega_2)$. Upon examining the pole-zero diagram depicted in figure 1.19, it becomes evident that this phase shift is positive; hence, the term 'lead compensator' is derived. The frequency at the maximum phase shift can be obtained by differentiating the phase angle, resulting in the following:

$$\omega_{max} = \sqrt{\omega_1\omega_2}, \qquad \theta_{max} = \sin^{-1}\left(\frac{\omega_2 - \omega_1}{\omega_2 + \omega_1}\right). \tag{1.26}$$

To achieve a substantial phase shift, it is necessary for the pole frequency (ω_2) to be significantly higher than the zero frequency (ω_1), with an upper limit of 90°. For the circuit illustrated in figure 1.18 (a), the relevant parameters can be calculated as follows:

- $\omega_1 = 10^4$ rad/s
- $\omega_2 = 2 \times 10^4$ rad/s
- $\omega_{max} = \sqrt{2} \times 10^4$ rad/s
- $\theta_{max} \approx 19.5°$.

These values exhibit reasonable agreement when compared to the simulation results presented in figure 1.18(b).

1.5.5 Lag compensator

The last circuit to be discussed in this section is the *lag compensator*, depicted in figure 1.20(a), with its corresponding Bode plots illustrated in figure 1.20(b). The transfer function can be derived using the same approach; here, Z_2 stands for the impedance of R_2 and C_1 in series:

$$H(s) = \frac{Z_2}{R_1 + Z_2} = \frac{R_2 + 1/sC_1}{R_1 + R_2 + 1/sC_1} = \frac{R_2}{R_1 + R_2}\frac{s + \omega_z}{s + \omega_p}. \tag{1.27}$$

In this equation, we have $\omega_z = 1/(R_2C_1)$, and $\omega_p = 1/[(R_1 + R_2)C_1)]$, leading to the relationship $\omega_z > \omega_p$. This indicates that the pole frequency is lower than the zero frequency, resulting in a negative overall phase angle of the transfer function, which gives rise to the terminology of *lag compensator*.

There are certain similarities with the lead compensator, particularly in terms of the expressions for the maximum phase shift:

$$\omega_{max} = \sqrt{\omega_z\omega_p}, \qquad \theta_{max} = -\sin^{-1}\left(\frac{\omega_z - \omega_p}{\omega_z + \omega_p}\right). \tag{1.28}$$

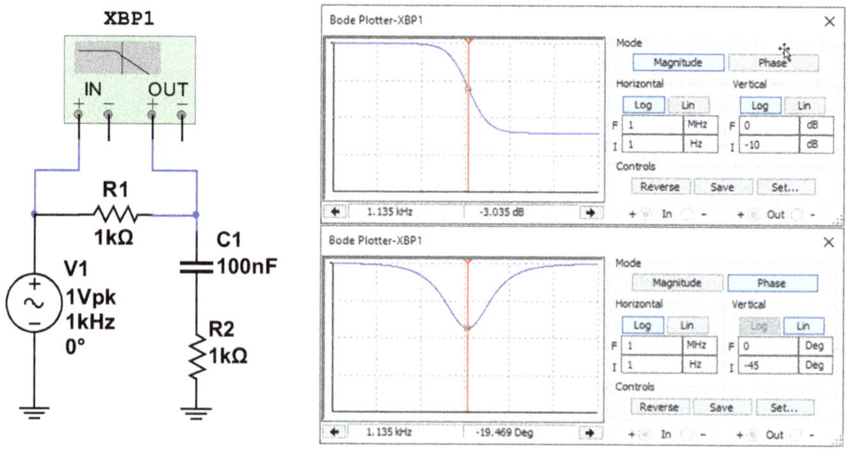

(a) (b)

Figure 1.20. Lag compensator: (a) circuit, (b) Bode plots.

Using the component values from the circuit illustrated in figure 1.20(a), the relevant parameters can be calculated:

- $\omega_z = 10^4$ rad/s
- $\omega_p = 5 \times 10^3$ rad/s
- $\omega_{max} \approx 7.07 \times 10^3$ rad/s
- $\theta_{max} \approx -19.5°$.

These calculated values demonstrate reasonable agreement when compared to the simulation results presented in figure 1.20(b).

Lead and lag compensator circuits are used to modify the frequency response of control systems and amplifiers. A lead compensator improves system stability and transient response by advancing the phase and increasing the BW. In contrast, a lag compensator enhances steady-state accuracy and reduces steady-state error by providing phase delay and attenuating high-frequency noise. These compensators are widely used in feedback and control applications to achieve the desired performance in terms of speed, stability, and accuracy.

1.6 Second-order LPFs and HPFs

Since the first-order LPF and HPF can be implemented with either a capacitor or an inductor, the second-order filters can be constructed with the combination of a capacitor and an inductor. However, as discussed in section 1.3, resonance could occur when these two devices are connected together. Figure 1.21(a) depicts a second-order LPF circuit, and its Bode plot is shown in figure 1.21(b).

First, we derive the general expression for the current signal in this circuit:

$$\tilde{I}(\omega) = \frac{\tilde{V}_S}{R + j\left(\omega L - \frac{1}{\omega C}\right)} = \frac{\tilde{V}_S}{R + j\omega L\left(1 - \frac{\omega_0^2}{\omega^2}\right)}. \tag{1.29}$$

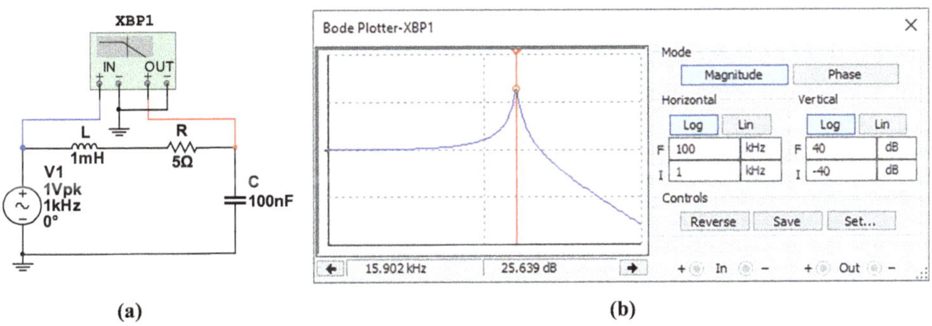

Figure 1.21. Second-order LPF: (a) circuit, (b) Bode plot.

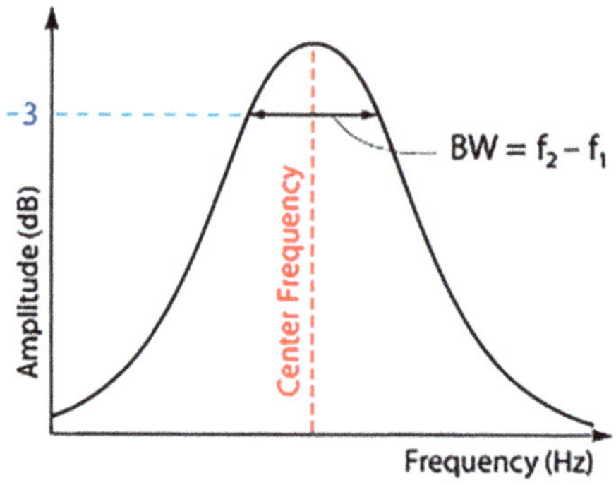

Figure 1.22. BW of a generic resonance peak. Created with GPT-4.0, OpenAI.

At the resonant frequency, $\omega_o = 1/\sqrt{LC}$, the impedances of the inductor and the capacitor nullify each other, causing the denominator to reach its minimum value. Consequently, the current exhibits a peak with zero phase shift at this frequency. On the other hand, as the frequency deviates from the resonant frequency, the balance between the inductor and the capacitor is broken, leading to a decrease in the current.

Figure 1.22 depicts a generic resonance peak, where the BW is defined as the frequency difference measured at the points where the amplitude drops 3 dB from the peak: $BW = \Delta f = f_2 - f_1$.

The BW of the resonance peak is inversely proportional to the quality factor Q of the RLC circuit, which can be calculated based on the values of the components: $Q = \sqrt{L/C}/R$. It is interesting to note that the BW does not depend on the value of the capacitor:

$$BW = \frac{f_0}{Q} = \frac{\omega_0}{2\pi Q} = \frac{1}{2\pi}\frac{R}{L} \tag{1.30}$$

This equation can also serve as the definition of the quality factor: $Q = f_0/BW$. In experimental setups, there are numerous parasitic circuit elements that are hard to identify and quantify. However, measuring the resonant frequency and the BW is straightforward, so this definition offers a practical approach for determining the quality factor.

1.6.1 Second-order LPF

The transfer function of an LPF, which is shown in figure 1.21(a), can be obtained by using the expression for current in equation (1.29), but it can also be directly derived from the equation for a voltage divider:

$$H_{\text{LPF2}}(\omega) = \frac{\widetilde{V}_c}{\widetilde{V}_s} = \frac{Z_C}{R + Z_L + Z_C} = \frac{1}{(1 - \omega^2 LC) + j\omega RC} = \frac{1}{\left(1 - \frac{\omega^2}{\omega_0^2}\right) + \frac{j}{Q}\frac{\omega}{\omega_0}}. \tag{1.31}$$

- At low frequencies, $\omega = \omega_0$, the transfer function becomes unity.
- At the resonant frequency, $\omega = \omega_0$, the transfer function is equal to $-jQ$, and there is a phase shift of $-90°$.

- At high frequencies, $\omega = \omega_0$, the transfer function can be approximated as $H(\omega) \approx -\omega_0^2/\omega^2$, decreasing at a rate of −40 dB dec^{-1} and having a phase shift of −180°.

The peak frequency can be determined by finding the derivative of the transfer function (magnitude) and setting it equal to zero, yielding a result slightly lower than the resonant frequency. Additionally, the peak value of the transfer function can also be found:

$$f_{\text{LPF, max}} = f_0\sqrt{1 - \frac{1}{2Q^2}}, \; |H|_{\text{max}} = \frac{Q}{\sqrt{1 - 1/(4Q^2)}}. \tag{1.32}$$

A peak or overshoot in the transfer function is highly undesirable in many applications. The condition for achieving a peak-free response can be derived from equation (1.32) by setting $|H|_{\text{max}} = 1$, and the result is $Q = 1/\sqrt{2}$. Therefore, if the quality factor is less than this threshold value, no overshoot occurs in the transfer function. In the case of the circuit depicted in figure 1.21(a), to meet this condition while keeping the capacitor and inductor unchanged, the resistor value should be increased to 141 Ω.

It might be initially assumed that the peak of this LPF should occur at the resonant frequency. However, clear understanding arises when the alternative method for deriving the transfer function is used—the voltage across the capacitor is equal to the product of the current and the impedance. As the magnitude of the impedance decreases with frequency, $|Z_C| = 1/\omega C$, it consequently shifts the peak frequency slightly lower.

1.6.2 Second-order HPF

If the positions of the capacitor and the inductor are swapped, the circuit becomes a second-order high-pass filter. Using the same procedure, the transfer function can be found:

$$H_{\text{HPF2}}(\omega) = \frac{\widetilde{V}_{\text{L}}}{\widetilde{V}_{\text{s}}} = \frac{Z_{\text{L}}}{R + Z_{\text{L}} + Z_{\text{C}}} = \frac{\omega^2 LC}{(\omega^2 LC - 1) - j\omega RC} = \frac{\omega^2}{(\omega^2 - \omega_0^2) - j\omega\omega_0/Q}. \tag{1.33}$$

- At low frequencies, $\omega = \omega_0$, the transfer function can be approximated as $H_{\text{LPF2}}(\omega) \approx -\omega^2/\omega_0^2$, with the magnitude increasing at at rate of 40 dB dec^{-1} and having a phase shift of 180°.
- At the resonant frequency, $\omega = \omega_0$, the transfer function is equal to jQ, and there is a phase shift of 90°.
- At high frequencies, $\omega = \omega_0$, the transfer function becomes unity.

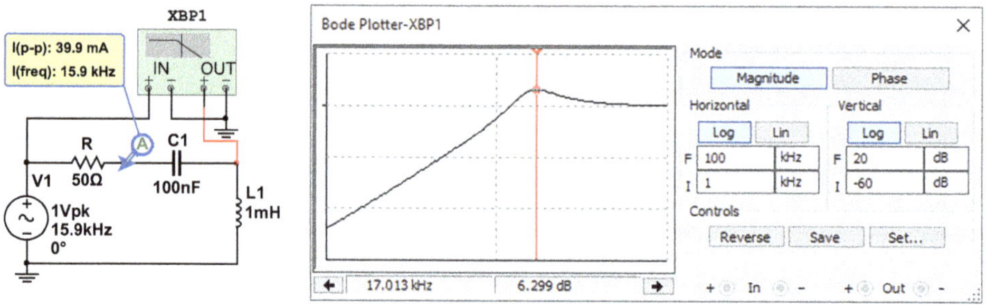

(a) (b)

Figure 1.23. Second-order HPF: (a) circuit, (b) Bode plot.

In addition, the peak frequency and the peak value of the transfer function can be determined via the approach used for the LPF:

$$f_{\mathrm{LPF,\ max}} = f_0\sqrt{\frac{1}{1 - 1/(2Q^2)}},\quad |H|_{\max} = \frac{Q}{\sqrt{1 - 1/(4Q^2)}}. \tag{1.34}$$

Unlike the LPF, the peak frequency of this transfer function is slightly higher than the resonant frequency due to the impedance of inductors increasing with frequency. However, the expression for the peak value of the transfer function remains the same as that of the LPF. Consequently, the condition for the quality factor without overshoot remains unchanged as well.

Figure 1.23 shows a second HPF circuit and its Bode plot. The resistance in the circuit is 50 Ω, resulting in a quality factor of $Q = 2$. At the resonant frequency, $f_0 = 15.9$ kHz, the capacitor and the inductor become nullified, resulting in a current amplitude of 20 mA. The peak-to-peak amplitude is shown in figure 1.23(a), which is about twice the amplitude. Applying the formulae in equation (1.34), the peak of the transfer function is at a high frequency, $f_{\max} \approx 17.0$ kHz, and the peak magnitude is at $|H|_{\max} \approx 2.066$, which can be converted to 6.30 dB. The simulation results shown in figure 1.23(b) agree with the calculation results very well.

> It is important to note that measuring the voltage across an inductor directly is not realistic, since there is always parasitic resistance associated with it. Consequently, the equations derived are applicable only for situations where a large series resistor is present in the circuit, which allows us to ignore the small parasitic resistance in the inductor.

1.7 First-order BPFs and BSFs

In addition to LPFs and HPFs, BPFs and BSFs are extensively utilized in electronics, and the magnitudes of their transfer functions are depicted in figure 1.24. BPFs enable the passage of signals within a specific frequency range while attenuating or blocking the signals with frequencies outside that range. They find

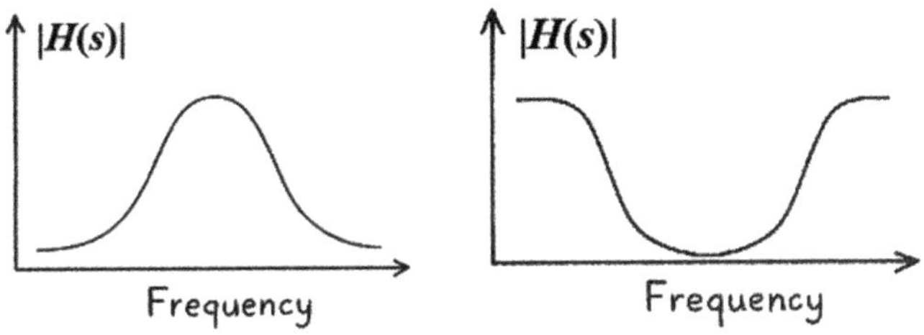

(a) (b)

Figure 1.24. Transfer characteristics of: (a) BPFs and (b) BSFs. Created with GPT-4.0, OpenAI.

Table 1.4. Behavior of series and parallel LC modules at the resonant frequency.

LC module	Impedance	Admittance	Behavior
Series	0	∞	Short circuit
Parallel	∞	0	Open circuit

applications in diverse fields where selective filtering of signals is essential. Conversely, BSFs, commonly referred to as notch filters, attenuate or block signals within a specific frequency range while allowing signals at all other frequencies to pass. They are widely employed in various fields where the suppression of specific frequency components is needed.

When a capacitor and an inductor are combined into a module, their impedance and admittance exhibit very interesting behaviors. There are two different ways of connecting them: series and parallel. For a series LC circuit, $Z_S = j[\omega L - 1/(\omega C)]$. For a parallel LC circuit, $Y_P = j[\omega C - 1/(\omega L)]$. Fortunately, the resonant frequency is the same, $\omega_0 = 1/\sqrt{LC}$, which was discussed in section 1.3. The behaviors of these two modules at the resonant frequency are summarized in table 1.4.

1.7.1 Series BPF circuit

The first BPF incorporates a series LC module, as illustrated in figure 1.25(a). The transfer function of this circuit can be obtained using the same derivation method.

$$H(s) = \frac{R_1}{Z_S + R_1} = \frac{R_1}{R_1 + j(\omega L_1 - 1/\omega C_1)} = \frac{1}{1 + jQ\frac{\omega}{\omega_0}\left(1 - \frac{\omega_0^2}{\omega^2}\right)} \tag{1.35}$$

- At the resonant frequency at $\omega_o = 1/\sqrt{L_1 C_1}$, the impedance of the series LC module vanishes, so the transfer function becomes unity. This resonance point is indicated by the cursor in the simulated Bode plots displayed in figure 1.25(b). It is worth noting that there is a slight discrepancy in the plot due to the inherent precision limitations of the simulation.

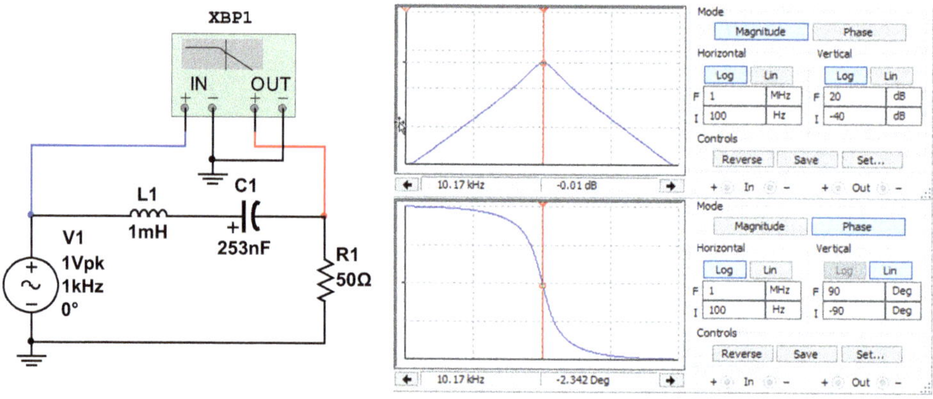

(a) (b)

Figure 1.25. BPF based on a series LC module: (a) circuit, (b) Bode plots.

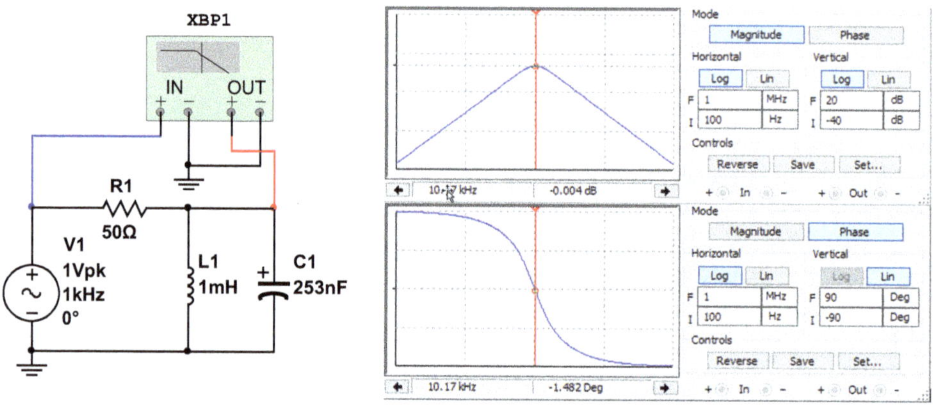

(a) (b)

Figure 1.26. BPF based on a parallel LC module: (a) circuit, (b) Bode plots.

– In the low-frequency domain, with the condition $\omega = \omega_0$, the capacitor becomes dominant, and the inductor can be neglected, resulting in the transfer function of $H(s) \approx j\omega R_1 C_1$. The curves on the left-hand side of the Bode plot can be approximated by this expression; for instance, the phase shift is 90°.
– In the high-frequency domain, with the condition $\omega = \omega_0$, the capacitor can be ignored, and the inductor plays the dominant role; thus, the transfer function becomes $H(s) \approx -jR_1/(\omega L_1)$. The curves on the right-hand side of the Bode plots are related to this expression, resulting in a phase shift of −90°.

1.7.2 Parallel BPF circuit

The second BPF incorporates a parallel LC module, as illustrated in figure 1.26(a), and the simulated Bode plots are presented in figure 1.26(b). The transfer function of this circuit can be derived in the same way.

$$H(s) = \frac{Z_P}{R_1 + Z_P} = \frac{1}{1 + R_1 Y_P} = \frac{1}{1 + jR_1(\omega C_1 - 1/\omega L_1)} = \frac{1}{1 + j\omega R_1 C_1(1 - \omega_0^2/\omega^2)} \tag{1.36}$$

- At the resonant frequency at $\omega_0 = 1/\sqrt{L_1 C_1}$, the admittance of the parallel LC module vanishes, so the transfer function becomes unity.
- In the low-frequency domain, with the condition $\omega = \omega_0$, the capacitor can be ignored, and the inductor takes on the dominant role; thus, the transfer function becomes $H(s) \approx j\omega L_1/R_1$, resulting in a phase shift of 90°.
- In the high-frequency domain, with the condition $\omega = \omega_0$, the capacitor becomes dominant, and the inductor can be neglected. This results in a transfer function of $H(s) \approx -j/(\omega R_1 C_1)$, and the phase shift becomes −90°.

1.7.3 Parallel BSF circuit

As discussed in section 1.6, an LPF circuit can be transformed into an HPF circuit by interchanging the positions of the resistor and capacitor or inductor. Similarly, if the positions of the LC module and the resistor are swapped, a BPF becomes a BSF. In the circuit depicted in figure 1.27(a), the parallel LC module acts like an open circuit at the resonant frequency, resulting in a deep dip in the Bode plots shown in figure 1.27(b). The transfer function of this circuit can be derived in the same way:

$$H(s) = \frac{R_1}{R_1 + Z_P} = \frac{R_1 Y_P}{1 + R_1 Y_P} = \frac{jR_1(\omega C_1 - 1/\omega L_1)}{1 + jR_1(\omega C_1 - 1/\omega L_1)} = \frac{j\omega R_1 C_1(1 - \omega_0^2/\omega^2)}{1 + j\omega R_1 C_1(1 - \omega_0^2/\omega^2)}. \tag{1.37}$$

Ideally, at the resonant frequency, the transfer function should approach zero without a phase shift, but the circuit simulator cannot handle this situation due to the limited precision of the numerical simulation. In practice, there is a noise floor in the output signal, making it unrealistic to reach signal powers below the noise level. On the other hand, at very low or very high frequencies, either the inductor or the

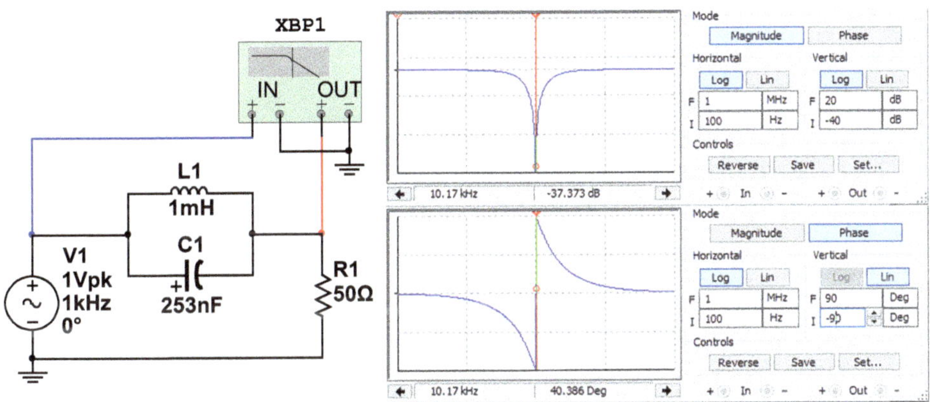

(a) (b)

Figure 1.27. BSF based on a parallel LC module: (a) circuit, (b) Bode plots.

capacitor behaves like a short circuit, causing the transfer function to approach unity, which is manifested in the Bode plots.

1.7.4 Series BSF circuit

Figure 1.28(a) displays the circuit of a BSF utilizing a series LC module, while its corresponding Bode plots are depicted in figure 1.28(b). Compared with the Bode plots in figure 1.27(b), it is evident that the frequency responses exhibit a striking similarity.

The analysis of this circuit is straightforward: at the resonant frequency, the series LC module behaves like a short circuit, causing the output signal to be significantly attenuated. Conversely, at very low or very high frequencies, either the capacitor or the inductor has a very high impedance and effectively becomes an open circuit, resulting in a unity transfer function.

1.8 Higher-order BPFs and BSFs

Among the four filter circuits discussed in the previous section, only a single LC module is utilized. It is worth noting that combining a series module and a parallel module can form a second-order filter, and the resulting circuits are illustrated in figure 1.29. The transfer functions of these circuits can be derived using the same

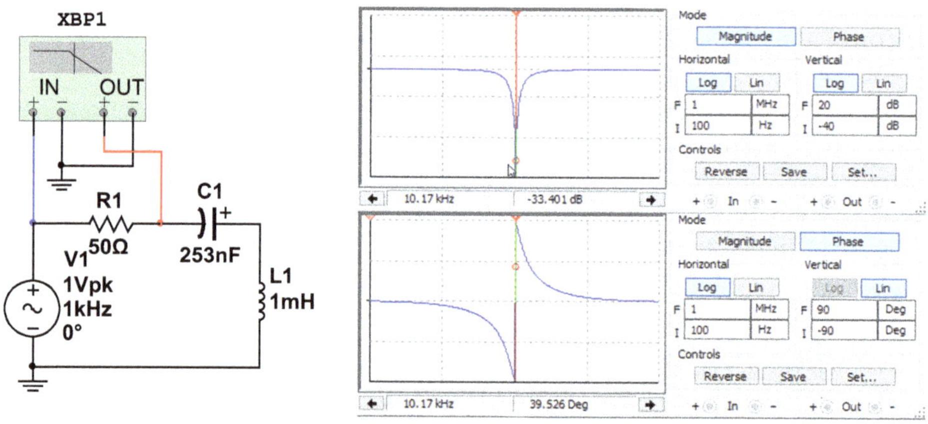

(a) (b)

Figure 1.28. BSF based on a series LC module: (a) circuit, (b) Bode plots.

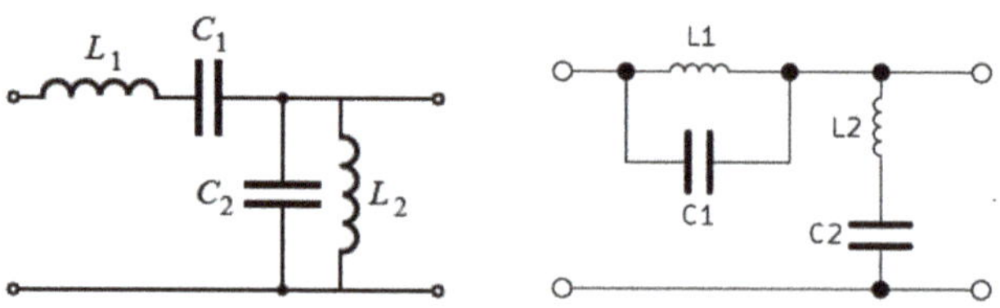

(a) (b)

Figure 1.29. Second-order filter circuits: (a) BPF, (b) BSF.

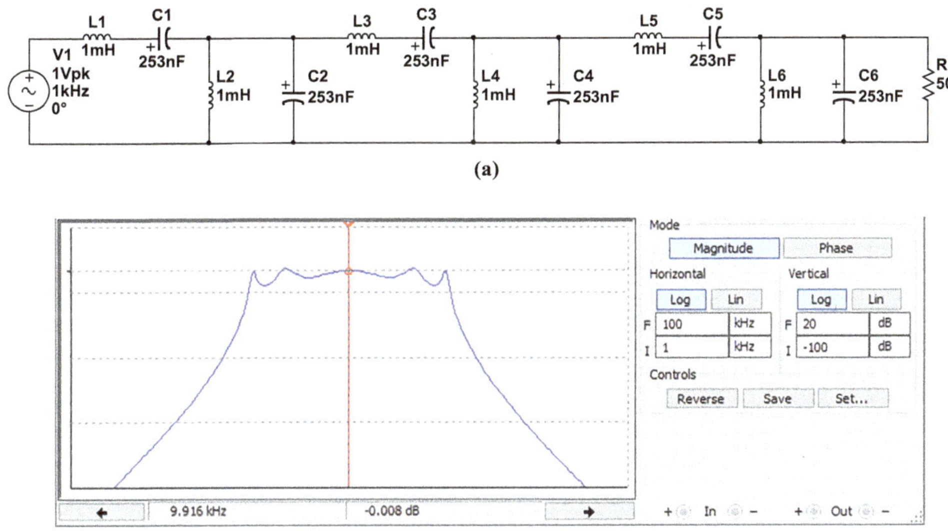

(a)

(b)

Figure 1.30. Sixth-order BPF: (a) circuit, (b) Bode plot.

methodology, and their behaviors can subsequently be determined based on the characteristics of these two modules. In addition, third-order filters are also very popular, and there are two different types of arrangements: T-type and Π-type.

In fact, the circuits depicted in figure 1.29 can serve as individual circuit modules, allowing multiple modules to be cascaded to create higher-order lattice filter circuits. Compared to lower-order filter circuits, these higher-order configurations offer improved performance. In addition, the device parameters of each module can be adjusted so that desirable results can be obtained.

A sixth-order BPF circuit is shown in figure 1.30(a), and its simulated Bode plot is illustrated in figure 1.30(b). Compared to the first-order BPF circuits shown in figures 1.25 and 1.26, the higher-order BPFs offer significant improvements: the peak in the pass band is transformed into a plateau, and the declines outside the edges of the pass band become much steeper.

> There are several well-known types of filters, including Butterworth filters, Chebyshev filters, and elliptic filters, each optimized for specific parameters. For example, Butterworth BPFs are celebrated for their capability to achieve a maximally flat response in the pass band, albeit with a less steep decline outside the pass band. On the other hand, Chebyshev filters offer a steeper roll-off at the expense of introducing ripples in the pass band. Meanwhile, elliptic BPFs can achieve the steepest roll-off, but they exhibit ripples in both the pass band and the stop band.

Similarly, higher-order BSFs can be constructed using the circuit module illustrated in figure 1.29(b). A sixth-order BSF circuit is shown in figure 1.31(a),

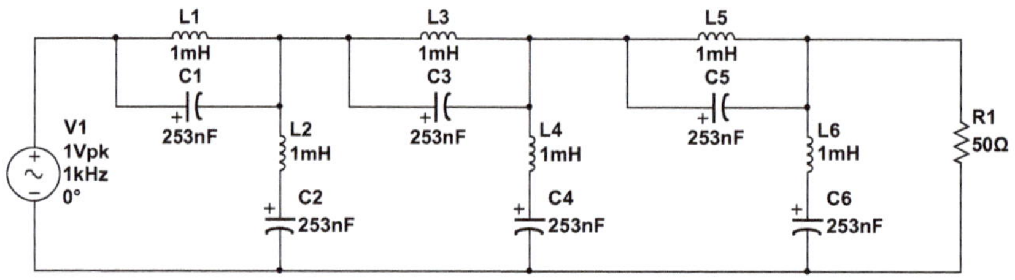

(a)

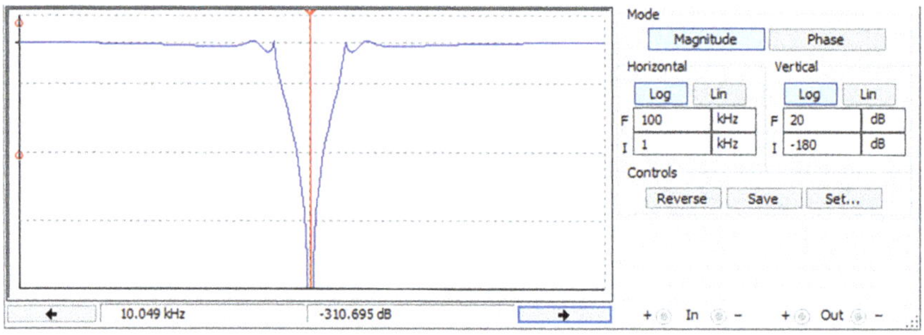

(b)

Figure 1.31. Sixth-order BSF: (a) circuit, (b) Bode plot.

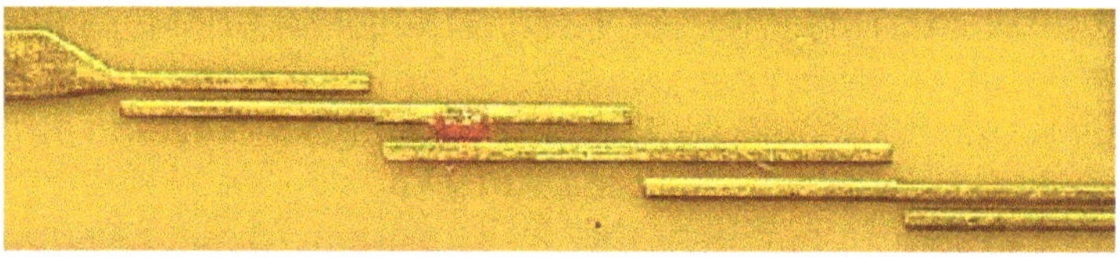

Figure 1.32. Coupled-line microstrip BPF. This [Microstrip-bandpass-filter] image has been obtained by the author from the Wikimedia website, where it is stated to have been released into the public domain. It is included within this book on that basis.

and its simulated Bode plot is depicted in figure 1.31(b). Compared to the first-order BSF circuits shown in figures 1.27 and 1.28, the Bode plots share the same feature, namely a notch filter, but the rejection in this circuit reaches more than −300 dB. In addition, the BW at −80 dB rejection is about 2 kHz, which is much broader than that of the first-order BSF circuits.

In addition to filters made using capacitors and inductors, microstrip filters can also be implemented on printed circuit boards for RF and microwave applications. An example of a microstrip BPF is shown in figure 1.32, where the long bars of copper play the role of inductors and the gaps between them form capacitors. In addition to this style of coupled-line filter, there are various other designs, such as hairpin and stub-loaded stepped-impedance filters.

Antireflection (AR) coatings in optics and electronic bandpass filters in circuits both achieve selective wave transmission through the use of layered or cascaded structures. In AR coatings, multiple thin-film layers with carefully engineered optical thicknesses create constructive and destructive interference that minimize reflections and enhance transmission at target band of wavelength. Likewise, higher-order electronic bandpass filters employ cascaded sections of capacitors, inductors, or microstrip resonators to expand the bandwidth of the passband. In both cases, adding more layers—whether optical films or filter stages—improves performance by expanding the bandwidth of transmitted waves.

Chapter 2

Semiconductor devices

Semiconductor devices play a crucial role in modern electronics and are the foundation of various electronic systems. They are used in applications ranging from basic electronic components to complex integrated circuits (ICs) found in computers, smartphones, telecommunications systems, and many other devices.

This chapter introduces the physical principles and operational mechanisms underlying modern semiconductor devices. Section 2.1 is an overview of energy band theory, which provides the framework for understanding how electrons and holes behave in solids. Section 2.2 covers carrier concentrations, including the role of doping in modifying electrical properties. The next two sections examine current transport mechanisms—drift and diffusion—that govern how charge carriers move in response to electric fields and concentration gradients. Building on these fundamentals, sections 2.5–2.7 analyze the behavior of the *pn* junction under equilibrium, reverse bias, and forward bias conditions, forming the basis for diodes. Diodes are two-terminal devices that allow current to flow in one direction while blocking it in the opposite direction. They are commonly used for rectification, signal modulation, and switching applications.

This chapter concludes with introductions to two essential active devices: the bipolar junction transistor (BJT) and the metal–oxide–semiconductor field-effect transistor (MOSFET), both of which are building blocks for analog and digital ICs. Unlike diodes, transistors are three-terminal devices that amplify or switch electronic signals and serve as the workhorse of analog and digital logic circuits. BJTs use both electrons and holes as charge carriers, while MOSFETs rely on only one type of carrier for the current.

2.1 Introduction to energy bands

Based on our understanding of atomic physics, the energy of electrons within atoms does not vary continuously but exists at discrete levels. The spatial distributions of electrons at different energy levels are figuratively called orbitals, which can be

doi:10.1088/978-0-7503-5512-4ch2

envisioned as satellites orbiting the Earth at different altitudes. While this analogy is intriguing, it is not entirely accurate, and a precise description necessitates the application of quantum mechanics.

The values of the atomic energy levels are strongly influenced by the atomic number, which corresponds to the number of protons and serves as the distinguishing factor between different elements. However, the arrangement of these energy levels remains consistent across all atoms, following a pattern such as 1s, 2s, 2p, 3s, 3p, 3d, and so on. Moreover, each orbital has a specific electron capacity, such as two electrons for an s-orbital and six electrons for a p-orbital, etc. Consequently, atoms with a greater number of electrons can occupy higher energy levels.

To illustrate the spatial distribution of these electrons, the onion model provides a simplified representation: electrons in lower orbitals are concentrated in layers closer to the nucleus, while those in higher energy levels occupy outer layers. The electrons located in the outermost layer are known as *valence* electrons, which play a significant role in determining the chemical behavior of an atom.

When two atoms are brought into close proximity, the core electrons remain unaffected. However, the orbitals of the valence electrons can overlap, allowing these electrons to travel between the two atoms. As a result, the energy levels of these shared electrons split into two distinct energy levels, as depicted in figure 2.1. Upon the formation of a stable bond, the lower energy level becomes occupied by electrons, while the higher energy level remains unoccupied.

Within a crystal structure, the phenomenon of orbital overlap extends across the entire material. Consequently, each atomic energy level of valence electrons broadens into a significantly wider energy band, comprising numerous closely spaced discrete energy levels. While the actual scenario is considerably more intricate, this simplified picture provides us with an intuitive comprehension of the energy band structure inherent to crystals.

Between atomic energy levels, there are wide energy gaps. Similarly, when the energy levels of the valence electrons expand into energy bands, usually there is also a gap between the bands, as illustrated in figure 2.2. This energy gap, commonly known as the *bandgap*, plays a pivotal role in determining various crucial properties of materials. For instance, if the bandgap is completely closed, the material behaves as a proficient conductor, such as a metal. Conversely, if the bandgap is exceptionally wide, the material acts as an insulator, such as quartz. Semiconductors lie between these two extremes. Furthermore, the bandgap profoundly influences the

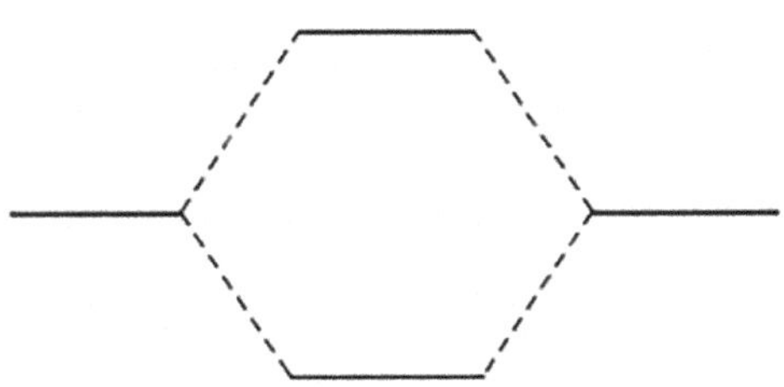

Figure 2.1. Split of an energy level. Created with GPT-4.0, OpenAI.

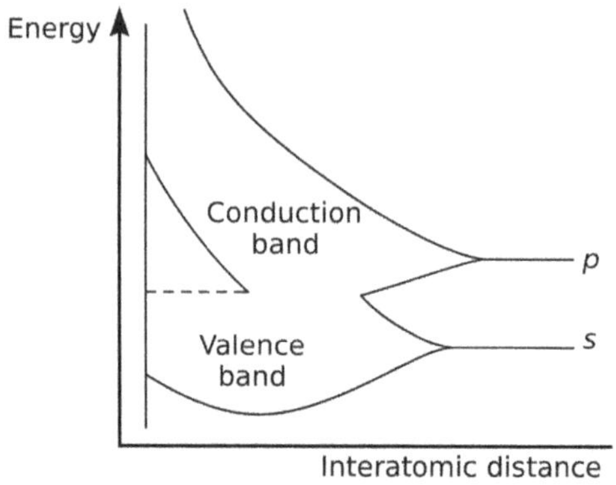

Figure 2.2. Transition from energy levels to energy bands. Created with GPT-4.0, OpenAI.

optical characteristics of a substance. Simply put, materials with wide bandgaps are transparent.

Silicon is the most important semiconductor in microelectronics. With an atomic number of 14, it contains 14 protons in its nucleus and is orbited by 14 electrons. Ten electrons populate the inner orbitals ($1s^2$, $2s^2$, $2p^6$), and the remaining four ($3s^2$, $3p^2$) are valence electrons. The silicon crystal used in microelectronics has a diamond lattice structure, where each atom forms symmetrical tetrahedral bonds with four neighboring atoms. This symmetrical arrangement causes the orbitals of the four valence electrons to undergo *hybridization*, resulting in the reorganization of the original atomic *s*- and *p*-orbitals and the emergence of four equivalent novel hybrid orbitals.

During the process of hybridization, certain aspects remain unchanged. For instance, in an isolated silicon atom, the 3s orbital contains two valence electrons, while a portion of the 3p orbitals accommodates two additional electrons, leaving four unoccupied spaces in the 3p orbitals. In essence, the 3s and 3p orbitals collectively have the capacity to house eight electrons, yet silicon atoms only utilize half of that capacity. Upon the formation of a silicon crystal, these hybridized orbitals segregate into two distinct groups of equal capacity, separated by a bandgap. Consequently, the group with lower energy (valence band) becomes fully occupied by electrons, while the group with higher energy (conduction band) remains largely unoccupied. This characteristic is prevalent in most intrinsic semiconductor materials, such as Ge, InP, GaAs, GaN, and others.

Figure 2.2 provides a visual representation of the transition from discrete atomic levels (on the right-hand side) to quasi-continuous energy bands (on the left-hand side) as the interatomic distance (horizontal axis) diminishes. Initially, as the atoms draw closer, the energy levels broaden, and the gap between them decreases. At a specific distance, this gap vanishes. However, as the atoms continue to approach each other, the gap reemerges and expands as the interatomic distance further decreases.

Intense pressure serves as one means to manipulate this distance, and it is employed in advanced research laboratories to investigate the band structures of

Table 2.1. Bandgap energies of selected materials at room temperature.

Semiconductor	Bandgap (eV)
Ge	0.67
Si	1.12
InP	1.35
GaAs	1.42
GaN	3.4
SiO_2	9

various materials. Another approach involves altering the temperature, as materials are known to expand with increasing temperature. This effect can be demonstrated simply by briefly submerging an LED in liquid nitrogen, which has a boiling point of 77 K. Such an experiment reveals a blue shift in the emitted light, indicating an increased bandgap. Table 2.1 provides the bandgap energy values for several widely used semiconductors and SiO_2 under standard atmospheric pressure and at room temperature.

The bandgap of a material plays a fundamental role in determining its optical properties, particularly its ability to absorb, transmit, and emit light. Only photons with energies equal to or greater than the bandgap can excite electrons from the valence band to the conduction band, resulting in the absorption of these photons. On the other hand, light can be emitted when electrons in the conduction band jump down to the valence band.

In wide-bandgap materials, such as GaN, the bandgap is large (e.g., >3 eV), meaning they are typically transparent to visible light. For example, materials such as quartz (SiO_2) fall into this category and can be used for optical windows. In contrast, narrow-bandgap semiconductors (e.g. Si and GaAs) have bandgaps below the visible spectrum (typically 1–2 eV) and are thus opaque to visible light. These materials strongly absorb light and are widely used in photodetectors and solar cells. On the other hand, the color of a light-emitting diode (LED) is directly related to the bandgap of its material.

At first glance, all intrinsic semiconductors should be insulators, since the conduction band is devoid of electrons, while the valence band is completely occupied. This scenario can be likened to a two-story classroom building with one large classroom on each floor: the classroom on the first floor is fully occupied while the one on the second floor remains entirely empty. Despite allowing students to switch seats, they are unable to do so due to the absence of available seats on the first floor. However, if some students are permitted to go upstairs, opportunities for movement are created. This is akin to how semiconductors facilitate the conduction of electricity.

Several factors contribute to the likelihood of electrons in the valence band transitioning to the conduction band. First, the bandgap energy plays a significant role in this process. In the case of materials with a large bandgap, such as quartz (SiO_2 crystal), few electrons possess the necessary energy to make the jump.

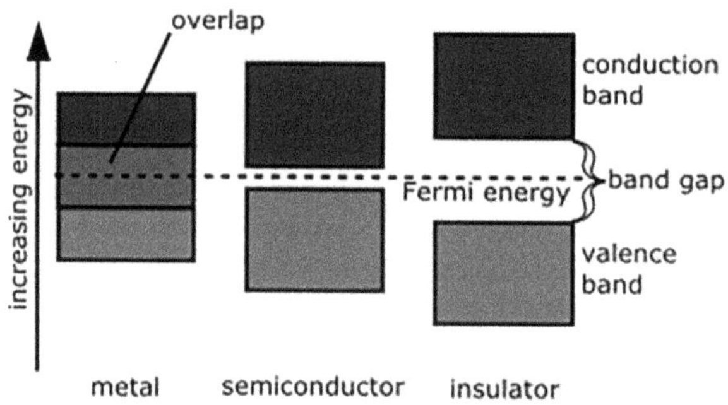

Figure 2.3. Relationship between conductivity and bandgap. This [Band gap comparison] image has been obtained by the author from the Wikimedia website where it was made available by [inductiveload] under a CC BY-SA 2.5 licence. It is included within this book on that basis. It is attributed to [inductiveload].

Consequently, SiO_2 becomes an effective insulator in electronic devices such as MOSFETs. Second, increasing the temperature allows more electrons to acquire sufficient energy for the transition. These excited electrons in the conduction band, along with the vacancies left behind in the valence band, can provide opportunities for electron motion. Hence, unlike metals, semiconductors exhibit a positive temperature coefficient in terms of conductivity, meaning they become more conductive at higher temperatures. While temperature can influence the bandgap to some extent, this effect is relatively weak and generally considered negligible.

Figure 2.3 illustrates the band structures of three classes of materials: metals, semiconductors, and insulators. In contrast to semiconductors and insulators, metals have no bandgap—either due to a partial overlap between the valence and conduction bands or because the conduction band is partially filled. As a result, a large number of electrons are free to move under the influence of an electric field. This high density of mobile charge carriers at the Fermi level accounts for the excellent electrical conductivity observed in metals.

> If the bandgap of semiconductors is not wide enough, the thermal stability of the devices becomes very poor. For instance, the first widely used semiconductor was germanium, with a bandgap of 0.67 eV at room temperature. Typically, its performance begins to deteriorate when its temperature reaches 70 °C. In contrast, silicon has a wider bandgap of 1.12 eV at room temperature, enabling well-designed silicon transistors to operate effectively up to 200 °C. In the field of power electronics, where elevated temperatures are common, semiconductors with wider bandgaps are preferred, such as silicon carbide (SiC).

2.2 Carrier concentrations

An intrinsic semiconductor is merely a theoretical model, and various impurity atoms are always present in real semiconductor materials. During the early history of the semiconductor industry, purification was an important advancement. For instance, silicon wafers with purity levels higher than 99.9% could be used for solar cells. However, the fabrication of ICs has much stricter requirements, and the impurity level needs to be reduced to below 10^{-6}. In other words, the purity level needs to be higher than 99.9999%. In the realm of semiconductors, the preferred unit of atomic concentration is the number of atoms per cubic centimeter (cm^{-3}). For example, the concentration of silicon atoms is about 5×10^{22} cm^{-3} and the impurity concentration in a blank silicon wafer is required to be below 10^{16} cm^{-3}.

In intrinsic semiconductors, the concentration of electrons in the conduction band is equal to the concentration of holes in the valence band. This equivalence arises because they are created simultaneously when electrons jump up from the valence band to the conduction band. As a result, these two concentrations can be expressed using a single parameter known as the *intrinsic carrier concentration* (n_i). The value of n_i is predominantly influenced by the bandgap (E_g) and temperature, and it can be calculated using the following equation:

$$n_i = \sqrt{N_C N_V} e^{-E_g/2kT}. \tag{2.1}$$

The product of the Boltzmann constant and absolute temperature forms a combination that is widely used in many situations. At room temperature, $kT \approx 25.9$ meV. At different temperatures, it can be calculated based on this value: $kT \approx 25.9\,(T/300)$ meV. The other two parameters, N_C and N_V, are the *effective density of states* in the conduction band and the valence band, respectively. Values of N_C, N_V, and the intrinsic carrier concentration are listed in table 2.2 for three different semiconductors at room temperature.

The effective densities of states of the three different semiconductors exhibit minor variations, but there is a notable disparity in the intrinsic carrier concentrations. This difference can be attributed to equation (2.1), where the dependence on the bandgap is an exponential function.

The effective density of states also changes with temperature, and it can be described by the following equation: $N(T) = N_{300}(T/300)^{1.5}$. This equation can be applied for both N_C and N_V, so the subscript is dropped. In addition, the parameter N_{300} stands for the value at $T = 300$ K, which is listed in table 2.2.

Table 2.2. Effective densities of states and intrinsic carrier concentrations of three different semiconductors at room temperature.

Parameters	Ge	Si	GaAs
N_C (cm^{-3})	1.05×10^{19}	2.82×10^{19}	4.37×10^{17}
N_V (cm^{-3})	3.92×10^{18}	1.83×10^{19}	8.68×10^{18}
n_i (cm^{-3})	2.3×10^{13}	1.0×10^{10}	2.1×10^{6}

When foreign impurity atoms replace host atoms in a semiconductor, they introduce energy levels within the bandgap. Some of these impurities generate energy levels positioned in the middle of the bandgap, referred to as *deep impurities*. Examples of deep impurities include various metal elements, such as Au, Cu, Mn, Cr, and Fe, among others, and their corresponding energy levels are illustrated in figure 2.4. These deep impurities are detrimental to the operation of most semiconductor devices and therefore need to be eliminated.

Conversely, certain impurity atoms create energy levels near the edges of the bandgap, known as shallow impurities, which play a valuable role in manipulating the properties of semiconductors. Shallow impurity atoms are intentionally doped into semiconductors, making this process a crucial step in the fabrication of ICs. Shallow impurities can be categorized into two types: *n*-type impurity atoms with five valence electrons (such as P and As) and *p*-type impurity atoms with three valence electrons (such as boron). Figure 2.5 shows their locations in the periodic table as well as the energy levels in the bandgap.

When an *n*-type impurity atom is introduced into a silicon crystal, four of its valence electrons form bonds with neighboring silicon atoms, while the remaining electron becomes free to move within the conduction band. This freely moving electron is referred to as a *carrier* because it can transport electric charge. Since electrons have a negative charge, they are known as *n*-type carriers. Furthermore, as these free-moving electrons originate from the impurity atoms, they are referred to as *donors*. On the other hand, when a *p*-type impurity atom is doped into a silicon crystal, it creates a vacancy or 'hole' in the valence band. Electrons from nearby regions can then move and occupy these vacancies, so they are called *acceptors*. Both types of doping can contribute to the movement of electrons and the conduction of current.

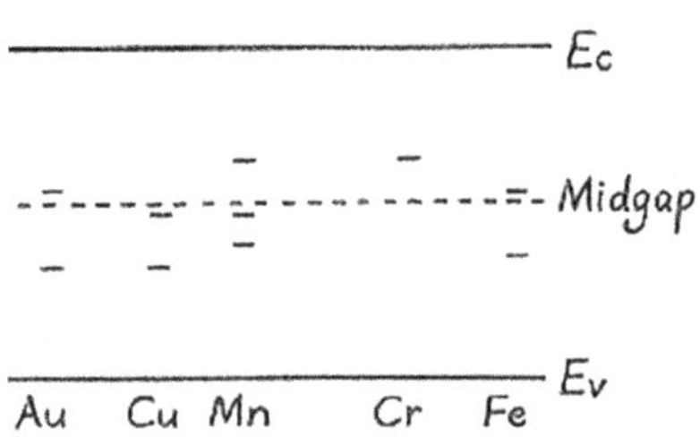

Figure 2.4. Energy levels of some deep impurity atoms. Created with GPT-4.0, OpenAI.

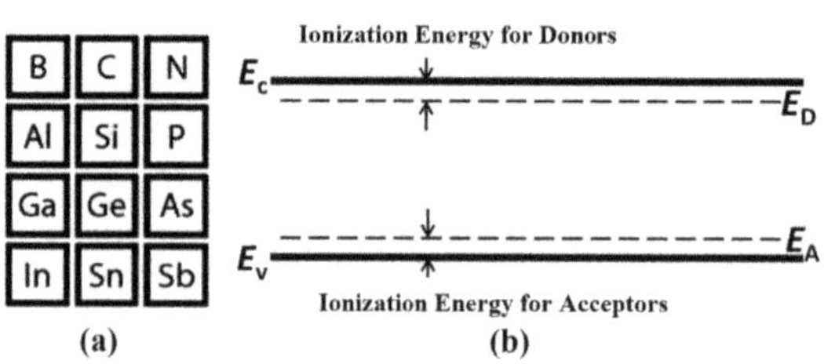

Figure 2.5. (a) Atoms in proximity to silicon, (b) energy levels due to shallow impurity atoms. Created with GPT-4.0, OpenAI.

At room temperature, it is a reasonable approximation to assume that all the shallow impurity atoms are *ionized.* This means that the additional electrons from the donor atoms are able to move freely in the conduction band, while the vacancies created by the acceptor atoms attract neighboring electrons to fill them. As a result, the concentration of majority carriers can easily be obtained: $n \approx N_D$ (for n-type doping), and $p \approx N_A$ (for p-type doping). In semiconductors, the concentrations of the minority and majority carriers are closely related, and they are inversely proportional. This means that their product remains constant, as indicated by the following equation:

$$np = n_i^2. \tag{2.2}$$

For example, when silicon is doped at $N_D = 10^{17}\ \text{cm}^{-3}$, it becomes straightforward to calculate the concentration of minority holes using this equation: $p = n_i^2/n \approx 10^3\ \text{cm}^{-3}$. It is important to note that in the context of semiconductors, the term 'minority' differs from its usage in social contexts, as the concentration disparity between minority and majority carriers is huge.

The derivation of equation (2.2) requires a certain level of understanding of statistical physics. Rigorously speaking, Fermi–Dirac statistics should be applied for electrons, but Boltzmann statistics are a good approximation in most situations. Under this approximation, the probability of occupation of a state with energy E can be expressed by a simple equation:

$$f_e(E) = \frac{1}{1 + \exp\left(\frac{E - E_F}{kT}\right)} \approx \exp\left(-\frac{E - E_F}{kT}\right). \tag{2.3}$$

In this equation, an important parameter is introduced, denoted by E_F, representing the *Fermi energy* or *Fermi level.* The condition for the validity of this approximation is $E - E_F > 3kT$. As discussed in the preceding section, both the conduction band and the valence band possess an enormous number of states. To better grasp this concept, an analogy can be employed. Visualize an expansive hotel comprising multiple floors, with each room symbolizing a quantum state within an energy band. Furthermore, each room can accommodate only one individual. Suppose the hotel manager intends to determine the total number of guests, and the crucial information lies in the occupation rates across each floor. Likewise, the distribution of states is well-established, allowing for the computation of electron or hole concentration through the utilization of the probability function.

If the elevators of the hotel are out of service, the situation resembles the conduction band, wherein the lower states have a higher probability of occupation, and equation (2.3) describes the occupation probability on different floors. However, the holes in the valence band are in the opposite situation: the top floors are very popular, with highly efficient elevators. As a result, the probability distribution for holes in the valence band acts as the complement to that of electrons:

$$f_h(E) = 1 - f_e(E) = \frac{\exp\left(\frac{E - E_F}{kT}\right)}{1 + \exp\left(\frac{E - E_F}{kT}\right)} = \frac{1}{1 + \exp\left(-\frac{E - E_F}{kT}\right)} \approx \exp\left(-\frac{E_F - E}{kT}\right). \tag{2.4}$$

In principle, calculating the concentration of electrons and holes using the distribution functions is possible. However, in practice, it poses a significant challenge due to the need to consider the contributions from all states. Therefore, a simplified approach is often preferred. By aggregating all states within the conduction band into a single energy level situated at the band edge, referred to as the effective density of states (N_C), a more manageable calculation is attained. Similarly, the valence band can be treated using the same approach, resulting in the effective density of states (N_V). This simplification streamlines the situation considerably, facilitating straightforward calculations for electron and hole concentrations:

$$\begin{aligned} n &= N_C \exp\left(-\frac{E_C - E_F}{kT}\right) \\ p &= N_V \exp\left(-\frac{E_F - E_V}{kT}\right). \end{aligned} \tag{2.5}$$

With these two equations, equation (2.2) can be proved by multiplying the concentrations of electrons and holes together, given the additional relationship: $E_g = E_C - E_V$. Moreover, it is important to note that equation (2.5) suggests that the carrier concentrations can be determined by the relative position of the Fermi level, assuming a fixed temperature. Figure 2.6 provides an illustration of the Fermi level positions under different conditions. Generally, as the doping level increases, the Fermi level moves closer to the band edge. In other words, in n-type semiconductors, the Fermi level rises towards E_C, but it drops toward E_V in p-type semiconductors. Furthermore, in extra highly doped semiconductors, the Fermi level may surpass the band edges. In such instances, the approximation of Boltzmann statistics becomes invalid, rendering the formulae in equation (2.5) inapplicable.

When the doping concentration remains constant, temperature variations can also induce a shift in the Fermi level. As an approximation, assuming N_C and N_V are constants, the trend of change can be determined using equation (2.5). With increasing temperature, the Fermi level gradually shifts towards the middle of the bandgap. Once a specific temperature threshold is surpassed, the intrinsic carrier concentration exceeds the doping concentration, resembling the behavior observed in intrinsic semiconductors.

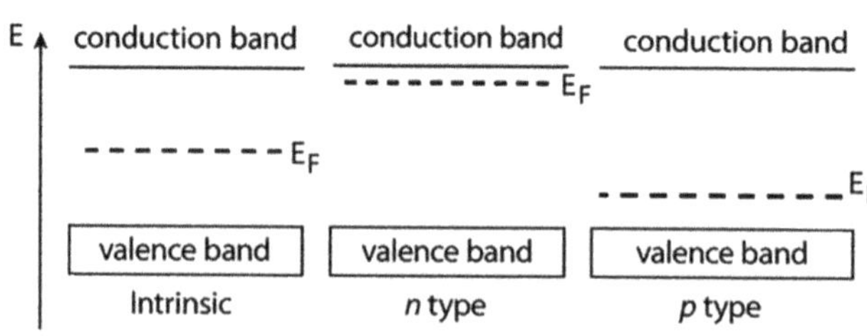

Figure 2.6. Relationship between Fermi level and doping. Created with GPT-4.0, OpenAI.

2.3 Drift current

An important property of carriers is their mobility, which is defined as the ratio of the drift velocity to the electric field. Equation (2.6) presents the simple relationship between these parameters. To aid in understanding, an analogy to skiing can be helpful: envision the electric field as the slope of a snow-covered mountain, the velocity as the speed of a skier, and the mobility as a measure of the slope's condition. The unit of mobility can be derived from this equation: $[\mu] = \text{cm}^2\ \text{V}^{-1}\ \text{s}^{-1}$.

$$v = \mu E \tag{2.6}$$

At first glance, this equation may seem counterintuitive. According to Newton's second law, the force derived from the electric field should be proportional to acceleration instead of velocity. The classical explanation draws an analogy to a densely forested slope where skiers frequently collide with trees, leading to an unpleasant experience at the ski resort. Similarly, the real scenario involves motion characterized by numerous short periods of acceleration rather than constant speed. Thus, the velocity in equation (2.6) can be interpreted as the average velocity, which relies on the magnitude of acceleration and the duration between successive collisions, known as the *free flight time*.

What parameters of semiconductors can be equated to the condition of the slope in a ski resort? One prominent parameter is the concentration of impurity atoms, whereby mobility decreases as the doping concentration increases. Additionally, the mobility of electrons differs from that of holes, with electrons generally exhibiting higher mobility. Figure 2.7 depicts the mobility of electrons and holes in silicon at various doping concentrations. It is evident that there is a noticeable decrease in mobility even at a modest doping level.

Temperature is another significant factor affecting mobility, as it induces lattice vibrations that generally lead to decreased mobility. At doping levels below 10^{17} cm^{-3}, a temperature increase of 100 °C can lead to a 50% drop in mobility.

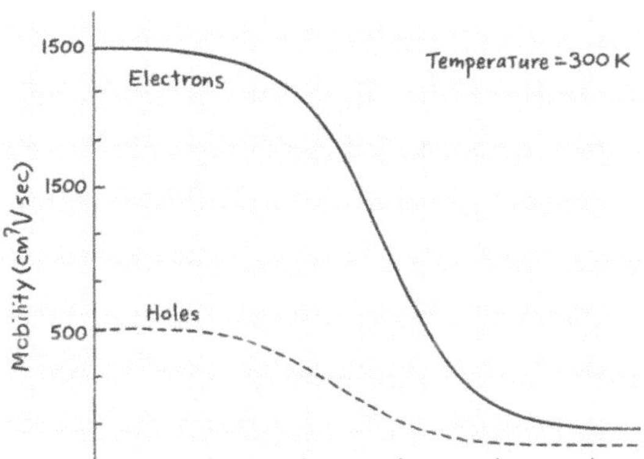

Figure 2.7. Electron and hole mobility vs. doping concentration. Created with GPT-4.0, OpenAI.

However, it is important to note the exception of high doping levels at low temperatures. When the doping concentration is greater than 10^{18} cm^{-3} and the temperature is below 300 K, mobility actually increases with temperature. This occurs because the heightened thermal energy at higher temperatures reduces scattering from impurity atoms, thereby enhancing mobility.

Table 2.3 presents the mobility values of three semiconductor materials at room temperature and at low doping levels. Although the mobilities of both electrons and holes are often listed with positive values, it is important to note that they move in opposite directions under an electric field, as holes are considered positively charged particles.

Based on the discussion above, it becomes evident that mobility lacks a firm foundation in physics and is defined as the slope of the velocity versus the electric field curve. Fortunately, in the low electric field range ($E < 10^3$ V cm^{-1}), this portion of the curve appears as a straight line for most semiconductor materials, as depicted in figure 2.8. However, as the electric field increases, the relationship becomes nonlinear. Notably, in the case of GaAs, the mobility of electrons starts to decrease beyond 3×10^3 V cm^{-1}. Furthermore, at very high electric fields, the drift velocities become saturated.

Table 2.3. Mobilities of electrons and holes at room temperature and at low doping levels.

Semiconductor	μ_e (cm^2 V^{-1} s^{-1})	μ_h (cm^2 V^{-1} s^{-1})
Si	1350	480
Ge	3900	1900
GaAs	8500	400

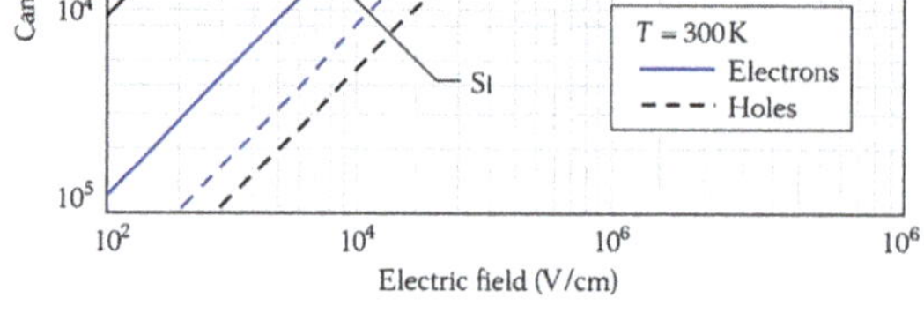

Figure 2.8. Mobility and velocity saturation. Created with GPT-4.0, OpenAI.

Another fundamental linear relationship associated with electric fields is the microscopic Ohm's law, represented by equation (2.7), where the parameter σ denotes conductivity in units of S cm^{-1}. This law is related to the more familiar macroscopic Ohm's law: $I = V/R$. Consider a uniform conducting rod with a length of L and a cross-sectional area of A. The current is related to the current density by $I = J \cdot A$, and the voltage is associated with the electric field by $V = E \cdot L$. By applying equation (2.7), the formula for resistance can be derived: $R = L/(\sigma A) = \rho L/A$. The parameter ρ stands for the resistivity of the material in units of $\Omega \cdot$ cm, which is the reciprocal of the conductivity:

$$\vec{J} = \sigma \vec{E}. \tag{2.7}$$

In equation (2.7), the conductivity σ is defined as the ratio between the current density and the electric field. It is determined by two factors: carrier density and mobility. To gain a better understanding of this relationship, let us take a step back and examine the expression for the current density from a microscopic perspective. For materials with a single type of carrier, the current density is proportional to the carrier density and the drift velocity:

$$\vec{J} = qn\vec{v}. \tag{2.8}$$

The carriers can be envisioned as vehicles transporting charges along a highway. To illustrate this analogy, imagine standing on a highway overpass and observing the vehicles traveling below. Assume the load (i.e. charges) is identical for all vehicles, which move with the same speed. In this case, the total charge passing through the overpass is proportional to the vehicle density and speed.

In semiconductors, both electrons and holes contribute to the current, so equation (2.8) should be modified to include two terms. By substituting equation (2.6) into the revised equation (2.8) and comparing it with equation (2.7), the expression for conductivity can be derived:

$$\sigma = e\left(\mu_{\mathrm{n}} n + \mu_{\mathrm{p}} p\right). \tag{2.9}$$

Since the concentrations of these two types of carriers are significantly different, the contribution from the minority carriers can be ignored. As a result, the dominant factor for conductivity is doping concentration, as the variation in mobility is relatively insignificant. Therefore, by measuring the conductivity or resistivity, the doping concentration of a wafer can be determined.

2.4 Diffusion current

When a teaspoon of juice is introduced into a cup of water, it rapidly disperses and eventually achieves uniform diffusion. In general, whenever there exists a disparity in concentration within a material, diffusion takes place. While we commonly observe this phenomenon in liquids and gases, it is important to note that diffusion also happens within solid materials, albeit often requiring higher temperatures to accelerate the process.

In the early years of the semiconductor industry, the doping process relied on diffusion techniques. For instance, triple-diffusion processes were employed in the fabrication of BJTs. One method of diffusion involved the use of a solid source. Initially, a layer of material containing either *n*-type or *p*-type dopant was deposited onto a silicon wafer. Subsequently, the wafer was placed into a diffusion chamber, where it underwent high-temperature treatment. As a result, the dopant atoms were activated by the heat and diffused extensively into the silicon wafer. In addition, gas-phase diffusion was also widely used for doping, which has the advantage of maintaining a constant dopant concentration at the boundary.

The diffusion process is governed by Fick's law. In a one-dimensional scenario, the simplified form in equation (2.10) can be employed. The diffusion coefficient, D, in units of $cm^2\ s^{-1}$, characterizes the ease with which a substance can diffuse within another substance, being notably high in fluids and very low in solids. The derivative term in the equation represents the gradient of a specific concentration. By definition, if the concentration $\phi(x)$ increases with x, its derivative is positive. However, it is important to note that diffusion always occurs from regions of high concentration to low concentration. Thus, a negative sign is incorporated into the equation to indicate the direction of diffusion:

$$J = -D\frac{d\phi}{dx}. \tag{2.10}$$

In semiconductors, the carrier concentration can exhibit spatial variations, and the diffusion of charged particles leads to electric current. The formula for the diffusion current density is very similar to Fick's law, with the inclusion of an additional factor for electric charge, given by $J = -qD(d\phi/dx)$. Specifically, $q = -e$ and $\phi = n$ for electrons, while $q = e$ and $\phi = p$ for holes, where e is the (positive) magnitude of the electron charge. The diffusion currents for electrons and holes are given by equations (2.11):

$$J_{\mathrm{n}} = eD_{\mathrm{n}}\frac{dn}{dx},\quad J_{\mathrm{p}} = -eD_{\mathrm{p}}\frac{dp}{dx}. \tag{2.11}$$

The contribution from minority carriers is equally significant in the case of diffusion current; in this respect, it differs from drift current. By considering both drift and diffusion currents, one can express the total current density for electrons and holes as follows:

$$\begin{aligned} J_{\mathrm{n}} &= e\mu_{\mathrm{n}}nE + eD_{\mathrm{n}}\frac{dn}{dx} \\ J_{\mathrm{p}} &= e\mu_{\mathrm{p}}pE - eD_{\mathrm{p}}\frac{dp}{dx}. \end{aligned} \tag{2.12}$$

There is a simple relationship between the diffusion coefficient and mobility, which is called the *Einstein relationship*:

$$\frac{D_{\mathrm{n}}}{\mu_{\mathrm{n}}} = \frac{D_{\mathrm{p}}}{\mu_{\mathrm{p}}} = \frac{kT}{e} = V_{\mathrm{T}}. \tag{2.13}$$

In the equation above, k represents the Boltzmann constant (1.38×10^{-23} J K^{-1}), kT is the *thermal energy*, and V_T is referred to as the *thermal voltage*. At room temperature, $V_T \approx 25.9$ mV, a value commonly used in the analysis of diodes and BJTs. Equation (2.13) implies that the diffusion coefficient of carriers is directly proportional to the mobility and temperature in kelvin. Nevertheless, because the mobility decreases with rising temperatures in most cases, the temperature dependence exhibits a sublinear behavior.

> In 1905, Albert Einstein made a significant contribution to the understanding of Brownian motion. His work on this phenomenon, conducted in the same year he discovered the theory of special relativity, provided a groundbreaking explanation for the erratic movement of small particles suspended in a fluid. By analyzing the random collisions between these particles and the surrounding fluid molecules, Einstein developed a mathematical framework that successfully explained the observed behavior of Brownian motion. This pioneering work not only shed light on the microscopic world but also demonstrated the profound connection between statistical physics and the behavior of particles on a molecular scale.

2.5 *pn* junctions at equilibrium

While certain devices, such as photoconductors, can be fabricated using semiconductors with a single type of doping, most devices require structures with both types of doping. This leads to the formation of a *pn* junction at the interface between these regions. Envision the formation of such a junction in real space. It can be visualized that the free-moving electrons in the *n*-type region diffuse into the *p*-type region, where they fill the available vacancies and become immobilized. From an energy band perspective, the electrons that migrate to the *p*-type region undergo recombination with the holes, resulting in their annihilation.

Nevertheless, this process cannot persist indefinitely, as a substantial electric field would rapidly develop. Prior to the contact between these two regions, both are electrically neutral. However, as electrons migrate across the interface, the *n*-type region in proximity to the interface becomes positively charged, while the *p*-type region on the opposite side acquires negative charges, as illustrated in figure 2.9. Consequently, when additional electrons attempt to migrate, they encounter a

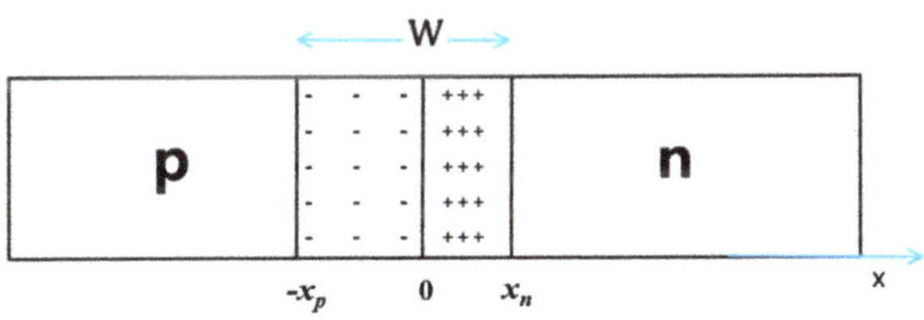

Figure 2.9. Space charge region in a *pn* junction. Created with GPT-4.0, OpenAI.

strong electric field that pulls them back into the n-type region. This phenomenon ensures that an equilibrium is ultimately established at the interface.

To simplify analysis, distinct boundaries are delineated between the neutral regions on both sides and the *space charge region* in the center. Furthermore, it is assumed that the doping concentrations remain constant on both sides, resulting in what is commonly referred to as a *step junction*. Since the concentrations of n-type donors (N_D) and p-type acceptors (N_A) typically differ, the widths of the space charge regions on both sides (x_p and x_n) vary accordingly. It is imperative to ensure the fulfillment of the charge neutrality condition:

$$N_A x_p = N_D x_n. \tag{2.14}$$

This equation reflects an idealized transient process for the formation of the pn junction: all the free-moving electrons in the space charge region on the n-type side migrate to the p-type side and effectively neutralize all the holes in the space charge region there. Consequently, no more carriers remain within the entire space charge region, earning it the name 'depletion region.'

In reality, an abrupt alteration in carrier concentration is unlikely to occur. Instead, a gradual transition takes place across the space charge region, where majority carriers gradually transform into minority carriers on the opposing side, as depicted in figure 2.10. Due to the significant change in carrier concentration over a short distance, the depletion model provides a useful approximation for estimating the width of this region.

In a pn junction, the existence of an electric field within the space charge region disrupts the alignment of the conduction band and the valence band. This scenario is illustrated in figure 2.11, where the displacement of the bands is dictated by the leveled Fermi energy, identified as E_F in the diagram. To draw an analogy, the Fermi energy can be likened to the concept of sea level, thus justifying its designation as the Fermi level. The criterion for achieving equilibrium is the constancy or flatness of the Fermi level in the band diagram.

Band diagrams typically depict the bandgap between the edges of the conduction and valence bands. However, it is important to remember that electrons can occupy a significant range above the conduction band edge, much as the atmosphere extends

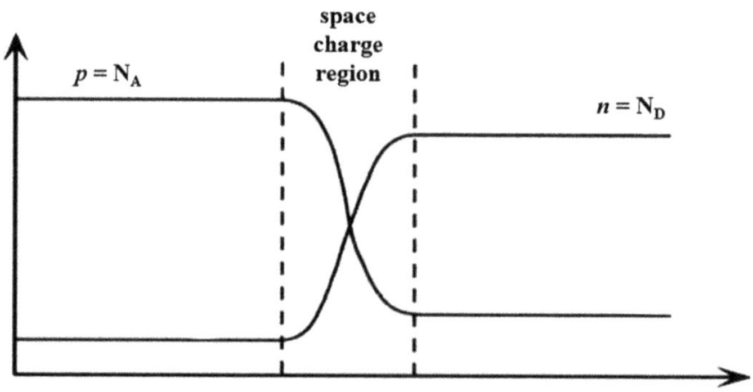

Figure 2.10. Carrier distribution in a *pn* junction. Created with GPT-4.0, OpenAI.

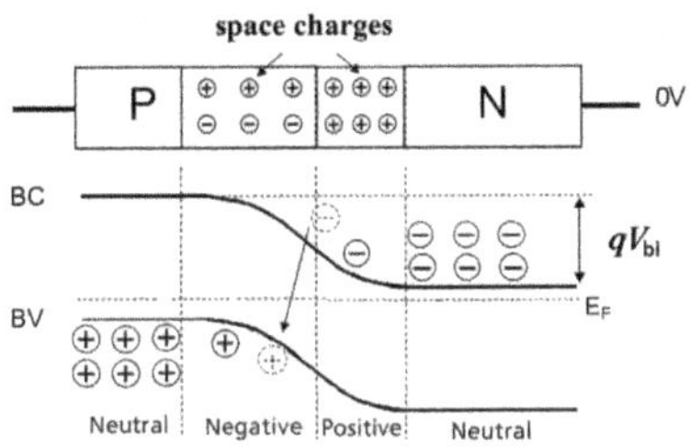

Figure 2.11. Band diagram of a *pn* junction, showing carrier distribution. Created with GPT-4.0, OpenAI.

far above the surface of the Earth. The same can be said for holes, which behave just like bubbles in water.

At equilibrium, there is no net current, but it is crucial to note that it is not a state of static inactivity. Instead, there exists a continuous process of generation and recombination of electron–hole pairs between the conduction band and the valence band. Figuratively speaking, electrons energetically jump up and down between the two bands, yet their overall effect is nullified.

There are a few important parameters for a *pn* junction, the first of which is the built-in voltage (V_{bi}), which is shown in figure 2.11. In the derivation of its expression, the Fermi level is used as the reference. As discussed in section 2.2, the relative position of the Fermi level is closely related to the doping concentration. For example, in the neutral region with *n*-type doping, the difference between the conduction band edge E_C and the Fermi energy can be expressed as $E_C - E_F = kT\ \ln(N_C/N_D)$. A similar equation is also available for the *p*-side, so the equation for the built-in voltage can be derived as follows:

$$V_{bi} = \frac{E_{C,p} - E_{C,n}}{e} = \frac{kT}{e}\ln\left(\frac{N_D N_A}{n_i^2}\right) = V_T\ \ln\left(\frac{N_D N_A}{n_i^2}\right) \tag{2.15}$$

As the doping concentration increases, the built-in voltage also increases, but its change is not very significant due to the logarithmic function. Typically, the Fermi energy resides within the bandgap, so this built-in voltage is usually lower in magnitude than the bandgap itself.

The next parameter to consider is the width of the depletion region (W), mainly determined by the doping concentrations, as shown in equation (2.16). Unlike the built-in voltage, this width is inversely proportional to the square root of the doping concentration, thus allowing for a wider range of variation. For example, a hundredfold increase in doping concentration results in a tenfold decrease in the width of the depletion region:

$$W = \sqrt{\frac{2\varepsilon_r \varepsilon_0 V_{bi}}{e}\left(\frac{1}{N_A} + \frac{1}{N_D}\right)}. \tag{2.16}$$

In the *pn* junctions of various semiconductor devices, a significant difference in doping levels exists between the two sides, resulting in a *one-sided junction.* In this configuration, the term associated with the higher doping concentration ($1/N_A$ or $1/N_D$) in equation (2.16) can be omitted, as the space charge region is predominantly located on the side with the lower doping concentration. Specifically, $W \approx x_p$ for a pn^+ junction, and $W \approx x_n$ for a p^+n junction.

The final parameter to be discussed is the peak electric field at the interface of a *pn* junction (E_{pk}), which can be expressed using the two previously derived parameters shown in equation (2.17). With an increase in doping concentrations, the built-in voltage experiences a slight increase, while the junction width decreases. Consequently, the peak electric field at the interface increases:

$$|E_{pk}| = \frac{2V_{bi}}{W}. \tag{2.17}$$

The peak electric field can be derived from Gauss's law, which takes into account the contributions from all the space charges on one side of the junction, as illustrated in figure 2.12. The sign or direction of this field is not inherently meaningful, as it would be reversed if the positions of the two regions were interchanged. Furthermore, integrating the electric field over the entire space charge region yields the built-in voltage, which is the area of the triangle at the bottom of figure 2.12.

> In the fabrication process of semiconductor devices, *compensated doping* is widely utilized. For example, the fabrication of a *pn* junction can start with a lightly doped *p*-type region, and then a portion of it is doped as *n*-type at a higher concentration. Consequently, this region is flipped from *p*-type to *n*-type, and the majority electron concentration can be approximated as $n \approx N_D - N_A$.

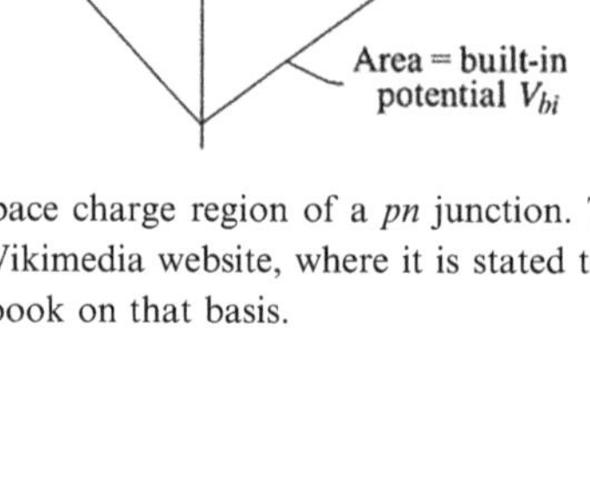

Figure 2.12. Electric field in the space charge region of a *pn* junction. This [PN overgang] image has been obtained by the author from the Wikimedia website, where it is stated to have been released into the public domain. It is included within this book on that basis.

2.6 *pn* junctions under reverse bias

Upon establishing ohmic contacts within the *p*-type and *n*-type regions, a diode is formed, as illustrated in figure 2.13(a). Diodes can be subjected to two distinct biases. In this section, we explore the scenario of reverse bias, while the next section delves into the case of forward bias.

When a moderate reverse bias is applied, the current flowing through the diode is extremely low, leading to the common assumption of negligible current in this situation. Nonetheless, a reverse bias does induce certain changes, including the expansion of the depletion region and an increase in the peak electric field. Interestingly, both parameters can be computed using essentially the same equation, albeit with a slight modification: $V_{\mathrm{bi}} \rightarrow V_{\mathrm{bi}} + V_{\mathrm{R}}$.

$$W = \sqrt{\frac{2\varepsilon_{\mathrm{r}}\varepsilon_0(V_{\mathrm{bi}} + V_{\mathrm{R}})}{e}\left(\frac{1}{N_{\mathrm{A}}} + \frac{1}{N_{\mathrm{D}}}\right)}, \qquad |E_{\mathrm{pk}}| = \frac{2(V_{\mathrm{bi}} + V_{\mathrm{R}})}{W} \tag{2.18}$$

If the reverse bias voltage is much higher than the built-in voltage, $V_{\mathrm{R}} = V_{\mathrm{bi}}$, the width of the depletion region and the peak electric field both become proportional to the square root of the bias voltage. For example, at $V_{\mathrm{R}} = 100\ V$, the width and the peak field are about ten times as large as in the equilibrium situation. This phenomenon can be experimentally verified by measuring the capacitance of the *pn* junction.

With the depletion region playing the role of an insulating layer, the structure of a *pn* junction can be viewed as a parallel plate capacitor, allowing for the straightforward determination of its capacitance:

$$C_{\mathrm{j}} = \varepsilon_{\mathrm{r}}\varepsilon_0\frac{A}{W} \Rightarrow \frac{1}{C_{\mathrm{j}}^2} = \frac{W^2}{(\varepsilon_{\mathrm{r}}\varepsilon_0 A)^2} \sim V_{\mathrm{bi}} + V_{\mathrm{R}}. \tag{2.19}$$

For data processing, a linear relationship is much easier to analyze. Hence, this equation can be reformulated into a new form, as shown in equation (2.19). Figure 2.14 illustrates this relationship, where the built-in voltage (V_{bi}) can be obtained by extending the plotted line to the left until it intersects the horizontal axis.

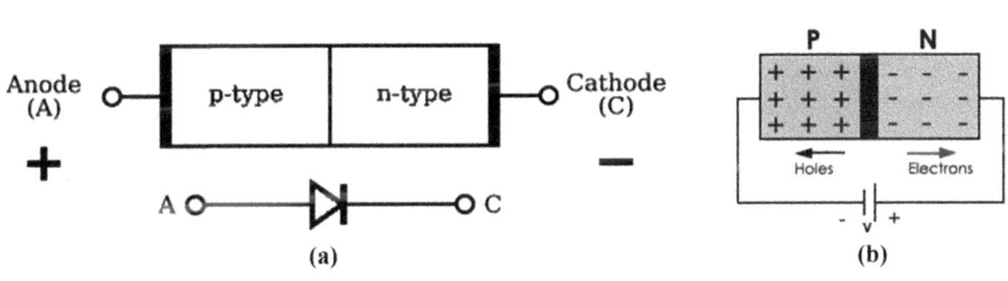

Figure 2.13. (a) A *pn* junction diode, (b) under reverse bias. This [PN Junction] image has been obtained by the author from the Wikimedia website, where it is stated to have been released into the public domain. It is included within this book on that basis.

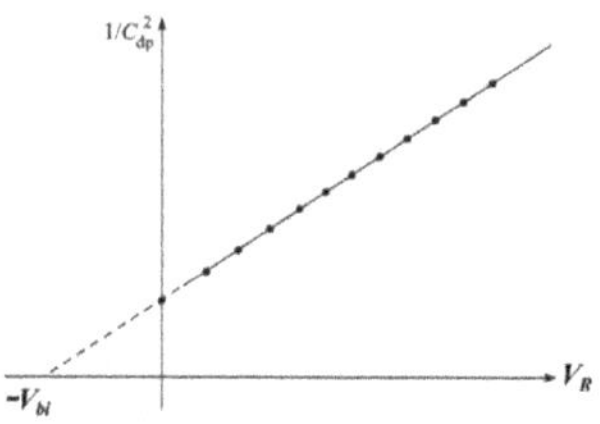

Figure 2.14. Capacitance of a *pn* junction under reverse bias. Created with GPT-4.0, OpenAI.

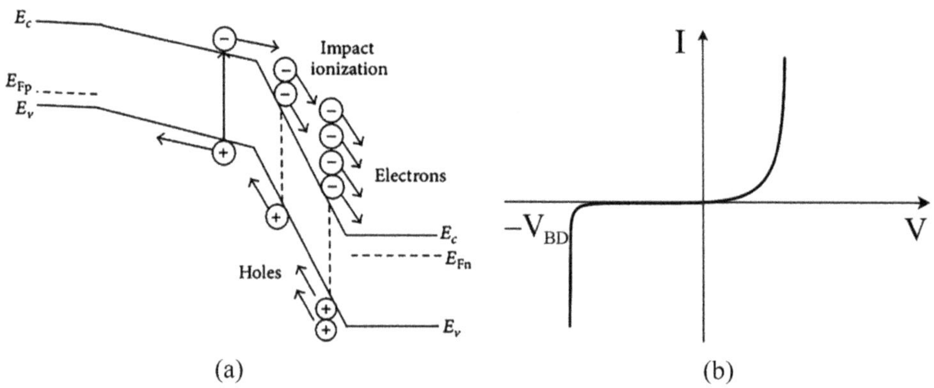

Figure 2.15. (a) Avalanche breakdown mechanism, (b) breakdown voltage and current. Created with GPT-4.0, OpenAI.

Leveraging the capability to modulate capacitance, a device known as a *varactor* can be created, which is a contraction of *variable capacitor*. Varactors find extensive utility in communication electronics circuits, such as voltage-controlled oscillators (VCOs).

One of the major applications of *pn* junction diodes is in rectifying circuits, where AC power is converted into DC power. For example, we need to recharge our cell phones regularly, and rectifier circuits are utilized in phone chargers. In such circuits, the diodes need to withstand a high reverse voltage, which is related to the peak electric field. In cases with exceedingly high voltage, avalanche breakdown can happen, which is illustrated in figure 2.15.

As explained in section 2.3, the electrons in semiconductors experience frequent collisions, limiting the distance they can travel freely. The average distance traveled between two consecutive collisions is called the *mean free path*. Its value primarily depends on doping concentration and temperature, which are the dominant scattering mechanisms. In moderately doped silicon at room temperature, the mean free path is in the range of a few nanometers to tens of nanometers. When subjected to a very strong electric field, an electron can gain enough energy in such a short distance to excite another electron from the valence band to the conduction band.

Under a sufficiently high reverse bias voltage, this process can resemble an avalanche, leading to a rapid increase in current, as shown in figure 2.15(b). Unfortunately, avalanche breakdown can potentially cause damage to diodes, so it should be avoided by choosing devices that can withstand sufficiently high voltages. Table 2.4 presents the specified peak reverse voltages for the diode models 1N4001–1N4007. As a general principle, reducing the doping concentration can increase the peak reverse voltage, offering greater resilience against avalanche breakdown.

When the doping concentration of a *pn* junction is high enough ($N > 5 \times 10^{17}\ \text{cm}^{-3}$), avalanche breakdown can also be avoided, as shown in figure 2.16. In this case, the electrons can tunnel directly from the valence band on the *p*-side to the conduction band on the *n*-side. Although the reverse current also increases rapidly, it does not cause damage to the diode. In other words, this is a normal working mode. In a well-designed Zener diode, the threshold voltage of this tunneling process is lower than that of the avalanche breakdown process, so the latter has no chance of being activated.

Zener diodes find extensive applications in voltage regulator circuits and also serve as crucial components for electrostatic discharge (ESD) protection. These diodes encompass a broad spectrum of threshold voltages, ranging from 2.4 to 1000 V, allowing for versatile usage across various voltage control and protection scenarios.

Table 2.4. Peak reserve voltages of rectifying diodes.

Parameter	1N4001	1N4002	1N4003	1N4004	1N4005	1N4006	1N4007
Peak reverse voltage (PRV)	50 V	100 V	200 V	400 V	600 V	800 V	1000 V
RMS reverse voltage	35 V	70 V	140 V	280 V	420 V	560 V	700 V

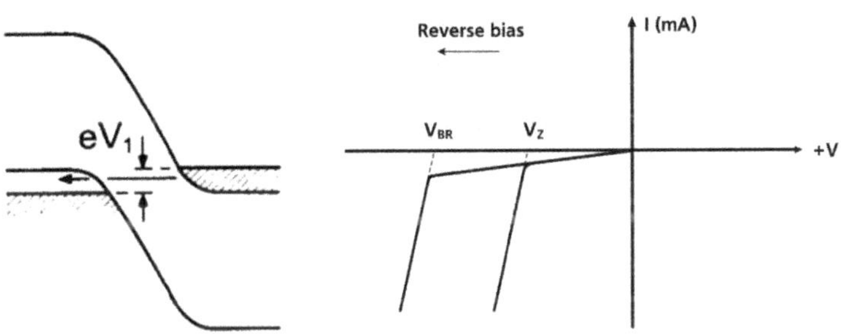

(a) (b)

Figure 2.16. (a) Electron tunneling process. (b) Zener and avalanche breakdown currents. Created with GPT-4.0, OpenAI.

2.7 *pn* junctions under forward bias

When a diode is subjected to forward bias, the current passing through it exhibits exponential growth, as depicted in figure 2.17(a). This phenomenon may sound familiar, as it is driven by the Boltzmann distribution of electrons in the conduction band and the corresponding mirrored distribution of holes in the valence band. With the *n*-side taken as a reference, the application of a forward bias voltage lowers the Fermi level on the *p*-side, resulting in an imbalance in electron and hole distributions, as illustrated in figure 2.17(b).

Drawing an analogy to the atmosphere, the *n*-type region resembles an expansive ocean, while the *p*-type region can be likened to a lofty plateau. The external bias assumes the role of adjusting the height of the plateau. When a reverse bias is applied, the plateau's elevation increases, causing the air to flow toward the ocean. However, due to the thinness of the air above the plateau, the resulting current is very weak. In contrast, under forward bias, the plateau's height is lowered suddenly, prompting the denser air above the ocean to flow inland. If we disregard the imbalances in the generation–recombination process within the space charge region, the resulting *I*–*V* curve can be succinctly described by a simple equation:

$$I_{\mathrm{D}} = I_{\mathrm{S}}\left[\exp\left(\frac{eV_{\mathrm{D}}}{kT}\right) - 1\right] = I_{\mathrm{S}}\left(e^{V_{\mathrm{D}}/V_{\mathrm{T}}} - 1\right). \tag{2.20}$$

When the forward bias voltage exceeds the thermal voltage V_{T} by a significant margin, the exponential term in the equation is much greater than one, so it can be further simplified to $I_{\mathrm{D}} = I_{\mathrm{S}}e^{V_{\mathrm{D}}/V_{\mathrm{T}}}$. Conversely, when a reverse bias voltage is applied, the exponential term is much less than one, resulting in a constant reverse current: $I_{\mathrm{D}} = -I_{\mathrm{S}}$. The parameter I_{S} earns its designation as the *reverse saturation current* based on the insights gained from this equation. The expression for I_{S} can be derived from the diffusion currents of minority carriers:

$$I_{\mathrm{S}} = eA\left[\left(\frac{n_0 D_{\mathrm{n}}}{L_{\mathrm{n}}}\right)_{p-side} + \left(\frac{p_0 D_{\mathrm{p}}}{L_{\mathrm{p}}}\right)_{n-side}\right] \tag{2.21}$$

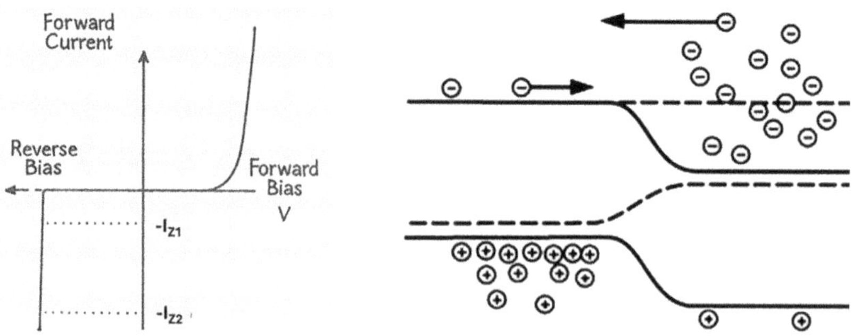

Figure 2.17. (a) *I*–*V* characteristics of a diode. (b) Carrier distribution under forward bias. Created with GPT-4.0, OpenAI.

- A: junction area
- L: diffusion length, $L = \sqrt{D\tau}$
- n_0 and p_0: equilibrium minority carrier concentrations, $n_0 = n_i^2/N_A$ and $p_0 = n_i^2/N_D$.

As we know, the intrinsic carrier concentration n_i is a strong function of temperature, thereby rendering the reverse saturation current also very sensitive to variations in temperature. Figure 2.18 illustrates the shift of the I–V curve in response to temperature change.

> The sensitivity of diode current is a challenge in electronic circuits. However, based on this effect, the semiconductor thermometer was invented. If the bias voltage of a diode is set at a constant level, the current passing through the diode changes with temperature, as shown in figure 2.18. Semiconductor thermometers have many advantages, including fast response, high accuracy, compactness, low power consumption, etc. Hence, they are widely used in various industries and home appliances.

When a weak low-frequency AC signal is superposed on a forward DC bias voltage, the diode exhibits AC behavior similar to that of a resistor. There are a few names for this behavior, including *AC resistance*, *incremental resistance*, and *differential resistance*. The last specifically pertains to its derivation from the expression of the I–V curve, and this resistance can be obtained by taking the reciprocal of the derivative of the current with respect to the bias voltage:

$$r_d = \left(\frac{dI_D}{dV_D}\right)^{-1} = \frac{V_T}{I_D}. \tag{2.22}$$

In a geometric interpretation, the derivative represents the slope of the tangent line at a specific point on a curve. For exponential curves, this slope increases with the

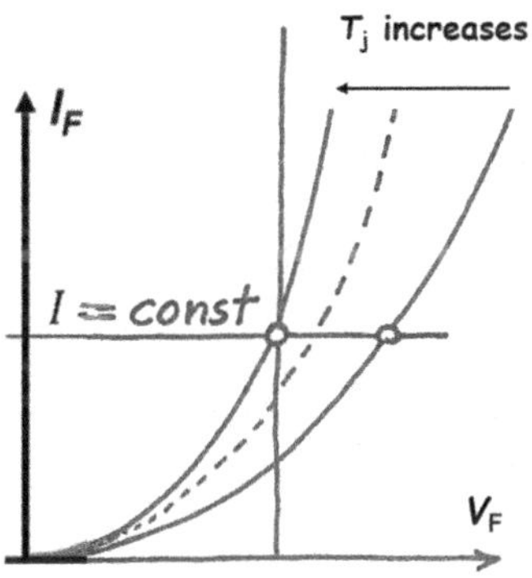

Figure 2.18. Temperature dependence of I–V characteristics. Created with GPT-4.0, OpenAI.

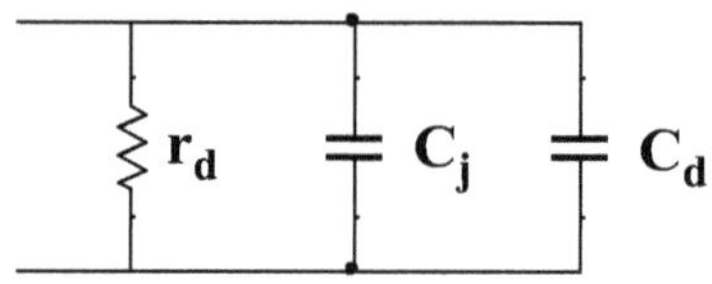

Figure 2.19. Small-signal model of a *pn* junction diode under forward bias.

forward bias voltage as well as the current. Equation (2.22) indicates that resistance decreases as the current increases. From this, a general conclusion can be drawn: certain AC parameters are determined by their corresponding DC parameters. In an analogy with stringed musical instruments, the AC signal resembles the vibration of the string, while the DC parameters correspond to the tuning of the string. Consequently, a specific point on the *I–V* curve is referred to as a *Q-point*, indicating a quiescent point or operating point.

As the frequency of the AC signal increases, the capacitive characteristics of a *pn* junction become more prominent. First, the junction capacitance (C_j) remains present, and its value increases as the width of the depletion region decreases under forward bias. Additionally, as discussed in section 2.5, the carrier concentration across the space charge region is a continuous function, and the applied AC voltage signal induces variations in the carrier distribution. This gives rise to another capacitance known as diffusion capacitance (C_d). In most situations, C_d is higher than C_j and thus plays the dominant role. Figure 2.19 illustrates a complete small-signal model of a forward-biased *pn* junction diode.

2.8 Bipolar junction transistors

In the early years of the twentieth century, two nascent industries found themselves in dire need of amplifiers: long-distance telephone systems and radio broadcasting. The realization of this dream came in 1907 with the invention of the thermionic triode, a vacuum tube that enabled signal amplification, albeit at the cost of being large and power-hungry. However, a breakthrough arrived four decades later with the invention of the BJT—a far more compact and efficient device. This transformative invention heralded an electronic revolution, sweeping the world with its wave of progress.

Bell Labs invented the BJT in 1947, and it follows the design principles of the triode vacuum tube: electrons are emitted from the emitter electrode and collected by the collector electrode, while a third electrode, known as the ***base*** due to its position at the bottom, modulates the current. Figure 2.20 illustrates the structure of an *npn* BJT. If the *n*-type and *p*-type regions are interchanged, it becomes a *pnp* BJT. Unlike a MOSFET, the BJT is an asymmetric device, featuring a thin, heavily doped emitter and a wide, lightly doped collector. The diagram denotes the difference in doping concentration with n^+ and n^- indicators.

Figure 2.21(a) depicts the band diagram of an *npn* BJT in equilibrium, exhibiting two *pn* junctions. A common question arises: can a BJT be considered as two *pn* junction diodes combined? The answer, in short, is ‘No.’ As electrons diffuse from the emitter to the collector, they encounter the *p*-type base region, where recombination

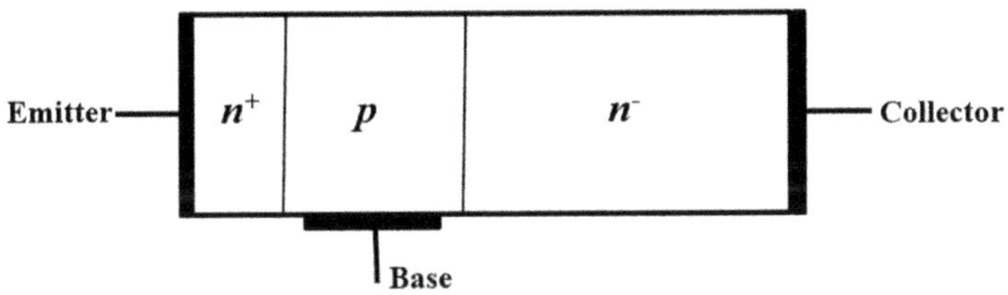

Figure 2.20. Structure of an *npn* BJT.

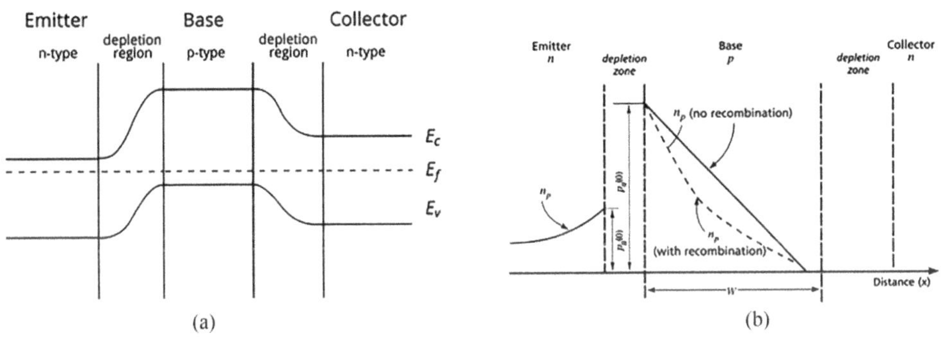

Figure 2.21. (a) Band diagram of an *npn* BJT. This [NPN Band Diagram Equilibrium] image has been obtained by the author from the Wikimedia website, where it is stated to have been released into the public domain. It is included within this book on that basis. (b) Minority carrier distribution. This [NPN BJT Minority Carrier Profiles (Active mode)] image has been obtained by the author from the Wikimedia website, where it is stated to have been released into the public domain. It is included within this book on that basis. Image enhanced by ChatGPT 5.0.

with holes can occur. This process is just like infantry soldiers passing through a minefield. To prevent this undesirable outcome of recombination, the base region needs to be very thin. However, if it becomes excessively thin, breakdown occurs at low voltages. In the early years of the semiconductor industry, this stringent requirement was a serious challenge in the development of the fabrication process.

Referring to the band diagram depicted in figure 2.21(a), the base region resembles a high barrier between the emitter and the collector, hindering the flow of electrons. Consequently, unless the barrier height is reduced by the application of a sufficiently high base voltage, the BJT is in the ***cutoff*** mode. It is important to note that the band diagram reflects the energy of negatively charged electrons, implying that a positive bias voltage lowers the energy level. In the case of a silicon BJT, the threshold voltage of V_{BE} begins at 0.5 V; however, a value of 0.7 V is necessary to achieve relatively high current, thus it is commonly used in DC analysis.

In amplifier circuits, BJTs typically operate in the ***active*** mode, where the emitter–base junction is forward biased, while the base–collector junction is usually reverse biased. Illustrated in figure 2.21(b) is the minority carrier distribution in this configuration, where the diffusion current of electrons in the base region plays a dominant role in the collection current for an *npn* BJT. Assuming negligible recombination with the holes, the electron distribution in the neutral base region

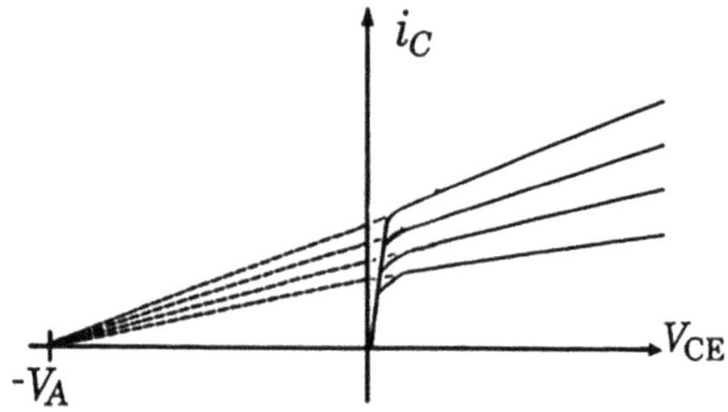

Figure 2.22. *I–V* characteristics of an *npn* BJT, showing the Early voltage. This [Early effect (graph - I_C vs V_{CE})] image has been obtained by the author from the Wikimedia website, where it is stated to have been released into the public domain. It is included within this book on that basis.

can be approximated by a linear function. As a result, a simple expression for the collector current can be derived:

$$I_C \approx eAD_{nB}\frac{dn}{dx} = \frac{eAD_{nB}n_{p0}}{w_B}(e^{V_{BE}/V_T} - 1). \tag{2.23}$$

In this equation, A stands for the junction area, while w_B denotes the *neutral* base width, which is the thickness of the base region with a flat band.

When the collector voltage increases, which plays the role of a reverse bias for the base–collector junction, the depletion region of this junction expands and the neutral base width (w_B) contracts, resulting in a slight increase in the collector current, as demonstrated in the *I–V* characteristics in figure 2.22. If the collector voltage reaches a very high level, causing the neutral base width to approach zero, breakdown occurs, leading to a rapid surge in the collector current.

As depicted in figure 2.22, when the lines in the active mode region are extended towards the left-hand side, they converge at a specific point on the horizontal axis. This crucial parameter of a BJT is known as the ***Early voltage*** (V_A), named in honor of James M Early (1922–2004), a distinguished American electrical engineer. For discrete transistors, the Early voltage typically exceeds 50 V. However, in ICs, their value tends to be significantly lower. A high Early voltage corresponds to flat *I–V* curves in the active region, resembling the behavior of a current source, which is ideal for an amplifier circuit. Conversely, a low Early voltage results in significant slopes of the curves within the active region, necessitating the inclusion of an output resistor in the small-signal circuit model.

In addition to the *cutoff* and *active* modes, a *saturation* mode exists within a narrow range on the left-hand side ($V_{CE} < 0.2$ V), where the collector currents display a rapid increase with V_{CE}, as shown in figure 2.22. In this mode, the conduction band in the base region is significantly lowered, causing a substantial influx of electrons to diffuse into the base region. However, if the conduction band in the collector region is not adequately lowered, these electrons cannot escape quickly, leading to their accumulation within the base region. The high concentration of electrons results in pronounced recombination with the holes present, leading to a high base current. This resembles the situation of rainwater accumulating on saturated ground, thus providing the origin of this terminology.

Using a similar approach to the derivation of equation (2.23), the base current can be approximated by the diffusion currents of the holes within the emitter region. Consequently, a similar equation can be obtained:

$$I_{\mathrm{B}} \approx -eAD_{\mathrm{pE}}\frac{\mathrm{d}p}{\mathrm{d}x} = \frac{eAD_{\mathrm{pE}}p_{\mathrm{n0}}}{w_{\mathrm{E}}}\left(\mathrm{e}^{V_{\mathrm{BE}}/V_{\mathrm{T}}} - 1\right). \tag{2.24}$$

The ratio of the collector current to the base current is the primary parameter of a BJT, which is called the *common-emitter (CE) current gain*:

$$\beta = \frac{I_{\mathrm{C}}}{I_{\mathrm{B}}} = \frac{D_{\mathrm{nB}}N_{\mathrm{D-E}}w_{\mathrm{E}}}{D_{\mathrm{pE}}N_{\mathrm{A-B}}w_{\mathrm{B}}} \tag{2.25}$$

In general, a higher current gain is more desirable, and equation (2.25) outlines specific requirements for BJT structures. First, it is essential that the doping concentration in the emitter exceeds that in the base, i.e. $N_{\mathrm{D-E}} = N_{\mathrm{A-B}}$. Second, the base width (w_{B}) should be significantly smaller than the emitter width (w_{E}). Finally, it is preferable for the diffusion coefficient of the holes in the emitter region (D_{pE}) to be low, which is why polycrystalline silicon is utilized in the emitter.

Figure 2.23 illustrates the current gain of the 2N3904 BJT at different temperatures and various current levels. It is well known that all semiconductor devices are highly susceptible to temperature changes, and this diagram demonstrates a notable increase in current gain with rising temperature. Additionally, the current gain decreases when the collector current becomes exceedingly high. However, when the collector current is within the range of 0.1–10 mA, a flat section is present, provided the temperature remains constant. Therefore, it is reasonable to assume that the current gain remains constant in circuit analysis.

In the datasheets of BJTs, the DC and AC current gains are commonly denoted by $h_{\mathrm{FE}} = \beta_{\mathrm{DC}} = I_{\mathrm{C}}/I_{\mathrm{B}}$ and $h_{\mathrm{fe}} = \beta_{\mathrm{AC}} = \frac{\partial i_{\mathrm{C}}}{\partial i_{\mathrm{B}}} = \frac{i_{\mathrm{c}}}{i_{\mathrm{b}}}$, respectively. It is important to note that these two parameters can differ if these two currents do not have a linear relationship. The letter *h* in these expressions stands for *hybrid parameter* in a two-port network, while the subscript *e* signifies the CE configuration in characterization, as shown in figure 2.24(a). In the hybrid configuration, the input signals

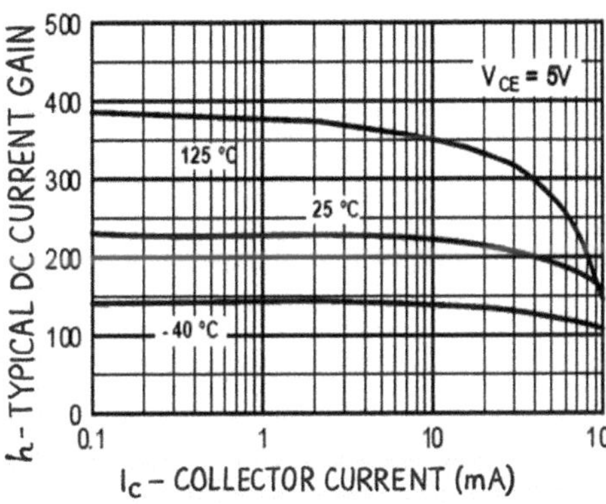

Figure 2.23. CE current gain of the 2N3904 BJT. Created with GPT-4.0, OpenAI.

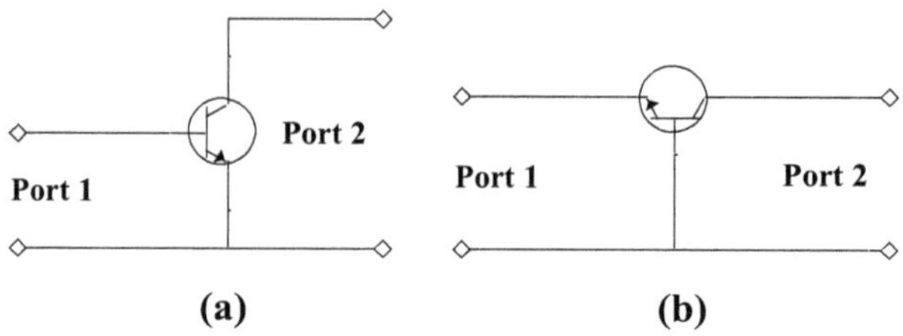

Figure 2.24. Two-port network diagrams for the (a) CE and (b) common-base configurations.

Table 2.5. Hybrid parameters of a BJT.

Meaning	BJT parameter	Two-port network
Input impedance	h_{ie}	$h_{11} = v_1/i_1$
Reverse feedback ratio	h_{re}	$h_{12} = v_1/v_2$
Forward current gain	h_{fe}	$h_{21} = i_2/i_1$
Output admittance	h_{oe}	$h_{22} = i_2/v_2$

comprise the base current (i_1) and collector-emitter voltage (v_2), whereas the output signals consist of the base-emitter voltage (v_1) and the collector current (i_2). In this way, the four hybrid parameters are listed in table 2.5.

In addition to the CE configuration, the triode vacuum tube-style common-base configuration was adopted to characterize the device in the early stages of BJT development, as depicted in figure 2.24(b). In this scenario, an alternative set of hybrid parameters can be defined, with two particularly useful ones being: $h_{ib} = \frac{v_1}{i_1} = \frac{v_{be}}{i_e} = r_e$ (a parameter used in the T-model; see section 3.5) and $h_{fb} = \frac{i_2}{i_1} = \frac{i_c}{-i_e} = -\alpha_{AC}$.

> As mentioned earlier in this section, the BJT was invented as a solid-state version of the triode vacuum tube. Consequently, the initial analysis of the BJT followed a similar approach to that of the vacuum tube. Therefore, the common-base current gain was represented by the first Greek letter, α. However, as the CE configuration gained more prominence, the second Greek letter, β, was used to denote the current gain in this configuration.

2.9 Metal–oxide–semiconductor field-effect transistors

The concept of the field-effect transistor (FET) was patented by Julius Lilienfeld in 1925, which predated the invention of the BJT by more than two decades. However, the MOSFET was invented at Bell Labs in 1959, a dozen years after the invention of the BJT. Initially, the MOSFET was referred to as an insulated-gate FET (IGFET), emphasizing its significant difference from the BJT—the insulation between the gate and the conducting channel. In the late 1970s, MOSFET ICs began to demonstrate advantages over BJTs in terms of high density and low cost, leading them to become the cornerstone of the microelectronics industry. While depletion-mode MOSFETs were widely employed in the early stages, they have now been largely replaced by enhancement-mode MOSFETs. As a result, this section solely focuses on the latter type of MOSFET.

Similar to the BJT, the MOSFET also encompasses two distinct types. The *n*-channel MOSFET (NMOS) bears a resemblance to the *npn* BJT, while the *p*-channel MOSFET (PMOS) shares similarities with the *pnp* BJT. The device structures of MOSFETs are illustrated in figure 2.25. In the central vertical cross-section, the device comprises three layers of different materials: a metal gate at the top, an oxide insulator beneath it, and the bulk silicon material—hence the acronym MOS, derived from the names of these materials. Conversely, along the horizontal direction, the structure is akin to that of an *npn* BJT. However, the MOSFET is a symmetric device, allowing for the interchangeability of the source and drain nodes.

Let us revisit the band diagram of an *npn* BJT depicted in figure 2.21(a), where the *p*-type silicon region creates a barrier to the movement of electrons. In the case of an NMOS, a similar situation arises, and current cannot flow unless the gate is appropriately biased. However, unlike the fixed threshold in BJTs, the MOSFET threshold voltage (V_{tn}) varies, depending on the parameters of the gate oxide layer and the doping level of the channel region. A criterion for this threshold voltage is depicted in figure 2.26(a), where the amount of band bending is equal to $2\phi_F$. This parameter, $\phi_{\text{F}} = E_{\text{i}} - E_{\text{F}}$, depends on the doping concentration of the channel region: $\phi_{\text{F}} = kT \ln(N_{\text{A}}/n_{\text{i}})$.

Similar to the *I–V* characteristics of an *npn* BJT, the curves illustrated in figure 2.26(b) can also be divided into three regions: *cutoff*, *triode*, and *active*. When the gate voltage is not sufficiently high ($V_{\text{GS}} < V_{\text{tn}}$), the current can be assumed to be zero, analogous to the *cutoff* mode in a BJT. If the gate voltage is sufficiently high ($V_{\text{GS}} > V_{\text{tn}}$), but the drain voltage is rather low, the current exhibits rapid changes with respect to the drain voltage. The shapes of these curves in this region resemble those of a triode vacuum tube, leading to its classification as the *triode* mode. Furthermore, when the drain voltage is very low, the *I–V* curves display

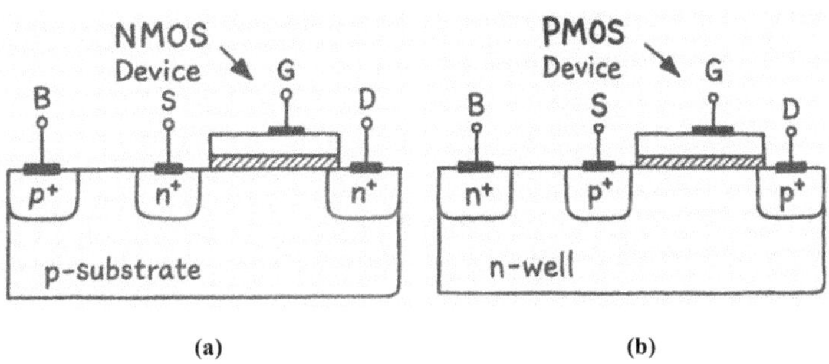

Figure 2.25. Device structures of traditional MOSFETs: (a) NMOS, (b) PMOS. Created with GPT-4.0, OpenAI.

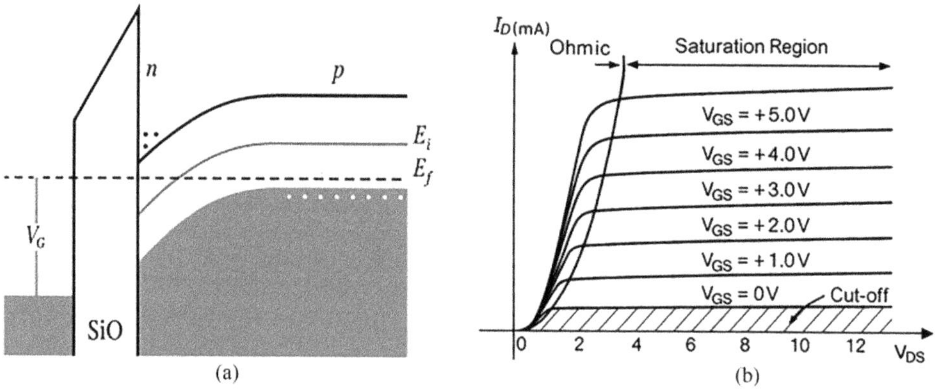

Figure 2.26. (a) Band diagram of an NMOS. This [MOS band bending] image has been obtained by the author from the Wikimedia website, where it is stated to have been released into the public domain. It is included within this book on that basis. (b) *I–V* characteristics. This [IvsV mosfet] image has been obtained by the author from the Wikimedia website where it was made available by [CyrilB~commonswiki] under a CC BY-SA 3.0 licence. It is included within this article on that basis. It is attributed to [CyrilB~commonswiki].

linearity, so the conducting channel behaves like a resistor with its resistance controlled by the gate voltage. For conventional NMOS, where the channel length is not extremely short, the current can be described by the following equation:

$$i_D = k_n\left[(v_{GS} - V_{tn}) - \frac{1}{2}v_{DS}\right]v_{DS} = k_n\left(v_{OD}v_{DS} - \frac{1}{2}v_{DS}^2\right) \tag{2.26}$$

In this equation, $k_n = \mu_n C_{ox} W/L$, μ_n, and C_{ox} are technology parameters, while W (width) and L (length) are the design parameters. Additionally, $v_{OD} = v_{GS} - V_{tn}$ is called the ***overdrive voltage***, which quantifies the extent to which the gate voltage surpasses the threshold voltage. When the drain voltage is extremely low ($V_{DS} = V_{OD}$), the second term in the equation becomes negligible. As a result, the *I–V* relationship assumes an ohmic behavior, and the channel resistance can be determined from this equation: $i_D = (k_n v_{OD})v_{DS}$.

When the drain voltage exceeds the overdrive voltage ($v_{DS} > v_{OD}$), its impact on the conducting channel diminishes, resulting in saturated curves. Hence, it is also referred to as the *saturation* mode. However, to avoid confusion with the saturation mode of the BJT, *active* mode is the preferred name for this region. The current can be approximated by a simple expression:

$$i_D \approx \frac{1}{2}k_n v_{OD}^2 \tag{2.27}$$

The boundary separating the triode mode and the active mode occurs at $v_{DS} = v_{OD}$, indicating that equations (2.26) and (2.27) should yield the same result at this point, which can be easily verified.

Just like BJTs, in the active mode, the drain-source voltage v_{DS} still exerts a weak influence on the current. As a *pn* junction exists between the *p*-type channel region and the *n*-type drain region, widening of the depletion region occurs when v_{DS} increases.

Consequently, the effective channel length becomes shorter, leading to a slight increase in the current. This effect can be described by the following equation, where the parameter λ is equivalent to the reciprocal of the Early voltage for a BJT: $\lambda = 1/V_A$.

$$i_D = \frac{1}{2}k_n v_{OD}^2(1 + \lambda v_{DS}) \tag{2.28}$$

> Unlike BJTs, MOSFETs are unipolar devices where either electrons or holes are solely involved in the transistor operation. Additionally, the current component resulting from minority carrier diffusion is significant in BJTs, while drift current predominates in MOSFETs. Generally, a drift process is more efficient than a diffusion process. Consequently, MOSFETs possess several advantages over BJTs. In digital circuits, MOSFETs outperform BJTs that rely on the slower diffusion of minority carriers, which can cause significant delays.

As predicted by Moore's law, transistors have continuously reduced in size, presenting a multitude of challenges. One such challenge is the drain-induced barrier lowering (DIBL) effect, which leads to significant leakage current even when the device is turned off by the gate voltage. Extensive research and development efforts have led to the adoption of a new MOSFET variant known as the FinFET, which has allowed Moore's law to remain valid for the transition from 28 nm down to 3 nm technology nodes. However, as the semiconductor industry progresses toward 2 nm technology and beyond, the MOSFET structure must undergo further modifications. To address this issue, gate-all-around (GAA) MOSFET technology has been implemented by Samsung in their 3 nm technology node for IC fabrication. The GAA MOSFET represents the next stage in the evolution of transistor design to meet the demands of advancing semiconductor technology. These three different device structures are depicted in figure 2.27.

> Moore's Law, named after Intel co-founder Gordon Moore, is the observation that the number of transistors on a chip doubles approximately every two years, leading to exponential growth in computing power. First stated in 1965, this trend has driven rapid advancements in semiconductor technology, enabling smaller, faster, and more energy-efficient electronic devices. Moore's Law has guided the semiconductor industry for decades, serving as a benchmark for innovation and miniaturization. While physical and economic challenges have slowed this pace in recent years, Moore's Law continues to influence research in new materials, architectures, and manufacturing techniques to sustain progress in computing performance.

In addition to their dominant role in ICs, MOSFETs are extensively utilized in power electronics applications. However, the structure of power MOSFETs differs to accommodate the flow of larger currents. Several alternative configurations exist,

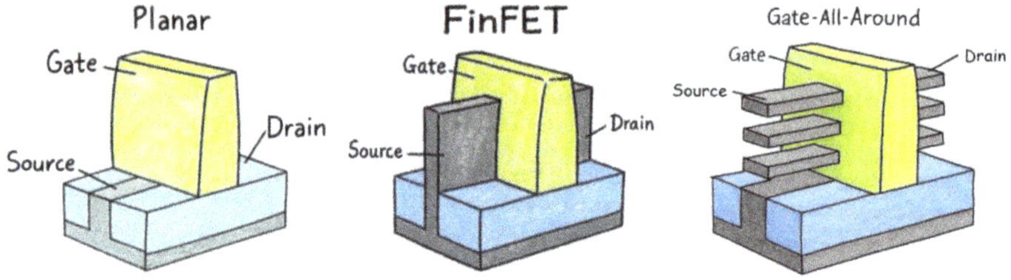

Figure 2.27. Three different MOSFET structures. Created with GPT-4.0, OpenAI.

including vertical MOS (V-MOS), depletion-mode MOS (D-MOS), and HEXFET configurations, among others. In these devices, the source and drain terminals are not interchangeable, necessitating clear identification when incorporating them into a circuit.

While MOSFETs offer the advantages of an insulated gate and the ability to handle larger currents compared to BJTs, they do have certain limitations. For instance, MOSFETs are not capable of withstanding extremely high voltages. To address this, a hybrid device known as the insulated-gate bipolar transistor (IGBT) has been developed. The IGBT combines the characteristics of both MOSFETs and BJTs, resulting in improved performance in power electronics applications when compared to either device individually.

The development of power transistors using wide-bandgap semiconductors such as silicon carbide (SiC) and gallium nitride (GaN) marks a significant advancement in power electronics. Unlike traditional silicon-based devices, SiC and GaN offer superior material properties, including higher breakdown voltages, greater thermal conductivity, and faster switching speeds. These characteristics enable power transistors to operate at higher voltages, frequencies, and temperatures, making them ideal for high-efficiency and high-power-density applications. SiC devices are particularly well-suited for high-voltage applications such as electric vehicles and industrial power systems, while GaN transistors excel in high-frequency, low-voltage environments such as RF amplifiers and power supplies. As manufacturing techniques mature and costs decrease, SiC and GaN devices are increasingly replacing silicon in performance-critical systems, offering substantial improvements in energy efficiency, size, and reliability. Their integration is driving innovation in renewable energy, automotive, aerospace, and consumer electronics, paving the way for more compact, efficient, and robust power conversion systems.

IOP Publishing

Essential Microelectronic Circuits (Second Edition)
A student's guide
Yumin Zhang

Chapter 3

Basic amplifier circuits

The early 20th century saw the electronics industry heavily focused on communication advancements. Wired telephones and wireless radio broadcasts were at the forefront, but both faced a common challenge: signal strength weakened with distance. To overcome this, amplifiers became the focus of the electronics industry. The invention of the vacuum tube in 1906, followed by its application in amplification circuits, revolutionized communication distances. This is evident in milestones such as the first transcontinental US phone call in 1915 and transatlantic phone calls made via submarine cables in 1927. Though the transistor arrived in 1947, surpassing the vacuum tube in performance, reliability, and cost, the groundwork for this advancement was laid by the earlier technology. This chapter delves into the fundamental configurations of amplifiers.

This chapter introduces the foundational principles and practical design of transistor-based amplifiers. It begins with the operating principles and transfer characteristics of transistor amplifiers, laying the groundwork for understanding signal amplification. The chapter then addresses the importance of establishing proper DC operating conditions and provides methods for designing effective biasing circuits. To support small-signal analysis, transistor models are introduced. The core sections focus on the common-emitter (CE) and common-source (CS) amplifier circuits, covering their behavior at low and high frequencies, as well as the impact of negative feedback. Finally, the chapter explores additional configurations, including common-base (CB)/common-gate (CG) and common-collector (CC)/common-drain (CD) amplifier circuits, rounding out a comprehensive overview of basic amplifier topologies.

3.1 Principle of transistor amplifiers

In the first two decades after the invention of the transistor, bipolar junction transistors (BJTs) were more popular than metal–oxide–semiconductor field-effect transistors (MOSFETs). However, when integrated circuits (ICs) were well developed and widely used, MOSFETs became dominant. First, they are more cost-

doi:10.1088/978-0-7503-5512-4ch3

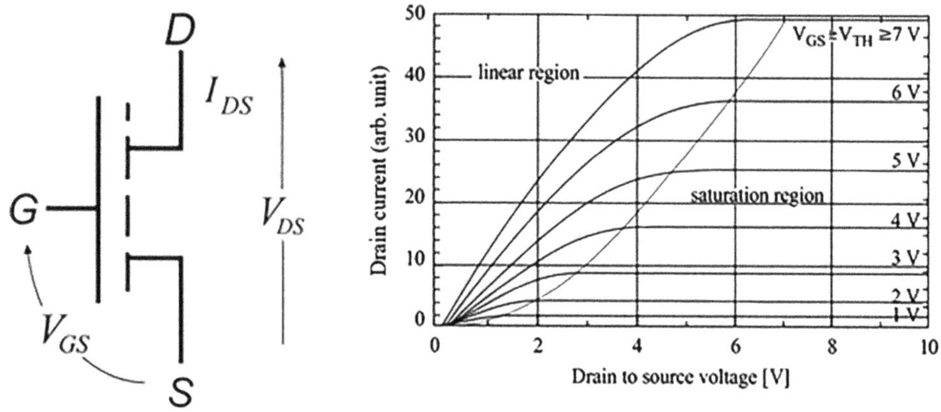

Figure 3.1. General *I–V* characteristics of a MOSFET. This [IvsV mosfet] image has been obtained by the author from the Wikimedia website where it was made available by [CyrilB~commonswiki] under a CC BY-SA 3.0 licence. It is included within this article on that basis. It is attributed to [CyrilB~commonswiki].

effective because the fabrication process of MOSFETs is simpler, and their density is higher. Second, the performance of MOSFET digital circuits is superior, while BJT circuits suffer from delays in the transition process. Figure 3.1 shows the typical *I–V* characteristics of an *n*-channel MOSFET, where the current is controlled by the two bias voltages: V_{GS} and V_{DS}. As discussed in the last section of chapter 2, the overdrive voltage V_{od} divides these curves into two regions: triode and active. In the active region, the effects of these two voltages on current differ significantly: the current changes markedly with V_{GS}, but it is not very sensitive to V_{DS}.

Since the drain current (I_D) remains relatively constant with changes in drain-to-source voltage (V_{DS}) in the active mode, this behavior is similar to that of a voltage-controlled current source (VCCS). This means it can effectively convert a voltage signal applied to the gate (v_{gs}) into a proportional current signal at the drain (i_d). The key parameter that determines this conversion efficiency is called *transconductance*:

$$g_m = \frac{i_d}{v_{gs}}. \tag{3.1}$$

Transconductance (g_m) is measured in Siemens (*S*), the reciprocal of resistance. It can also be expressed in other forms, such as mho, which is ohm spelled backward. Figure 3.2(a) shows an amplifier circuit using a VCCS with a transconductance of 0.1 Mho. For a sinusoidal input voltage signal with an amplitude of 1 mV, this VCCS would produce an output current signal with an amplitude of 0.1 mA. The resistor above the VCCS converts this current signal back into a voltage signal. This demonstrates the basic principle of a transistor amplifier: one stage converts a voltage signal into a current signal, and another stage converts the current signal back into a voltage signal. The gain of the amplifier is determined by the properties of these stages:

$$|A_V| = g_m R. \tag{3.2}$$

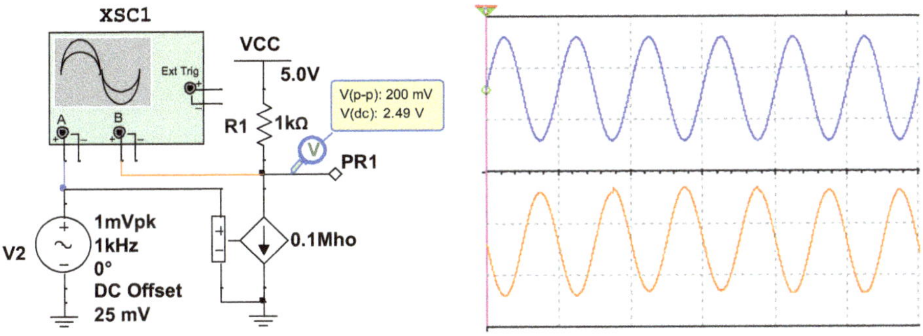

(a) (b)

Figure 3.2. Amplifier with voltage-controlled current source: (a) circuit, (b) simulated waveform.

Using the parameters provided for this circuit, the gain can be found easily: $|A_V| = 100\,(\mathrm{V/V})$. The simulated result in figure 3.2(a) verifies this: the peak-to-peak output voltage is 200 mV, so its amplitude is 100 mV, which is 100 times the input signal's amplitude. However, from the simulated waveform shown in figure 3.2 (b), we find that there is a phase reversal. Mathematically, the gain becomes negative due to this 180-degree phase shift between the input and output signals: $A_V = -100\,(\mathrm{V/V})$. This does not affect the amplifier's ability to boost the signal strength and is not considered a limitation.

> Electronic amplifiers typically work using a two-step process. In the first step, a weak voltage signal is converted into a current signal. In the second step, this current signal is converted back into a voltage signal with a greater magnitude. However, it is important to note that there are other types of amplifiers that function differently. For instance, optical amplifiers directly amplify the intensity of light in just one step.

Analyzing amplifier circuits mathematically requires considering two sets of parameters: DC and AC. By convention, DC parameters are denoted by uppercase letters with uppercase subscripts (e.g. V_{CC}, I_C). AC parameters, on the other hand, are represented by lowercase letters with lowercase subscripts (e.g., v_{in}, i_c). A hybrid notation using lowercase letters with uppercase subscripts (e.g., v_E, i_C) indicates a combination of DC and AC parameters. In figure 3.2(a), both the DC voltage and the amplitude of the AC signal are shown at the output node. We can analyze these components separately using Ohm's law:

$$
\begin{aligned}
V_O &= V_{CC} - I_O R_1 = V_{CC} - g_m R_1 V_{IN} \\
v_o(t) &= 0 - i_o(t) R_1 = -g_m R_1 v_{in}(t) \\
v_O(t) &= V_O + v_o(t) = V_O - g_m R_1 v_{in}(t).
\end{aligned}
\tag{3.3}
$$

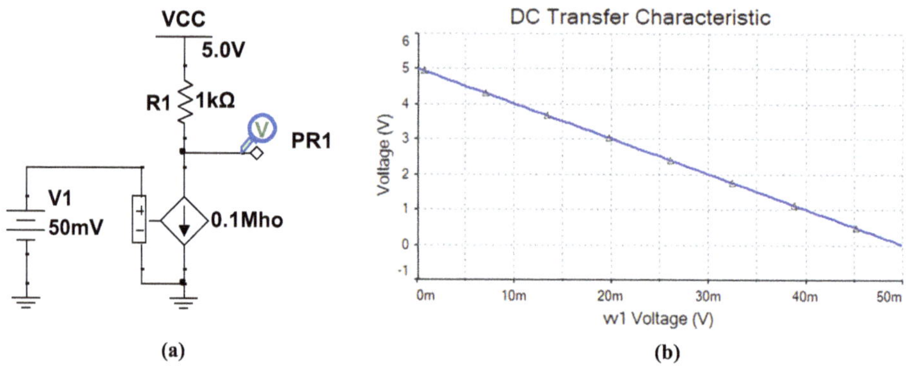

Figure 3.3. Transfer characteristics: (a) circuit, (b) simulation with DC sweep.

Oscilloscopes offer two coupling modes to display signals: DC coupling and AC coupling. DC coupling displays the entire signal, including both its DC offset and the AC signal. However, if the DC component is much larger than the AC component, it is very challenging to display the AC signal on the screen, which drifts far away from the level in the middle. In such cases, AC coupling is preferred. This acts like the addition of a high-pass filter to the signal, thus removing the DC component and centering the AC waveform on the oscilloscope's display.

Amplifiers can also be characterized through their input–output transfer characteristics. In figure 3.3(a), the input signal is replaced by a DC voltage source, and its value is swept over a range using the 'DC Sweep' simulation mode of Multisim. The simulation result is illustrated in figure 3.3(b), showing the output voltage (vertical) as a function of the input voltage V1 (horizontal). The gain can be calculated from the slope of this curve, which is equal to -100 V/V. However, for transistor amplifiers, this linear relationship between input and output only holds for a very limited range of input voltages. In other words, transistors behave like a current source only under well-designed DC bias conditions.

3.2 Transfer characteristics of transistor amplifiers

Since the I–V characteristics of a transistor in active mode resemble those of a VCCS, we can substitute the VCCS in figure 3.3 with a transistor, as shown in figure 3.4. The simulated transfer characteristics of this BJT amplifier exhibit a descending slope only within a very narrow range (V1 $\sim$ 0.66–0.71 V). This behavior differs significantly from the straight line observed in figure 3.3(b). The amplifier's gain, calculated from the slope (dy/dx) of the transfer curve in figures 3.4(b), is -82.2 V/V, which is shown in the inset of figure 3.4(b).

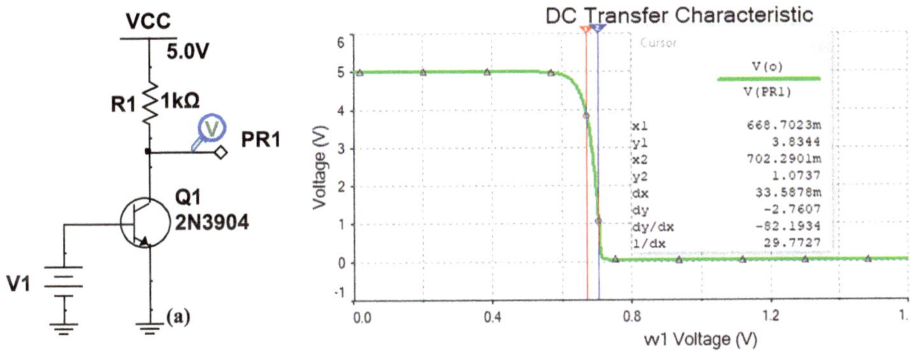

(a) (b)

Figure 3.4. Transfer characteristics of a BJT amplifier: (a) circuit, (b) simulation with DC sweep.

The behavior on either side of this narrow transition region is particularly interesting for digital logic circuits. For low input voltages (V1 < 0.65 V), the output voltage remains high (around 5 V). Conversely, for high input voltages (V1 > 0.71 V), the output goes low (around 0 V). This characteristic suggests the potential application as a digital logic device—an inverter or NOT gate. For such an application, a resistor is needed between the voltage source and the base of the BJT for protection.

From another perspective, the transistor also works as a 'switch.' With a high input voltage (V1 > 0.71 V), the switch is closed, and the transistor acts as a short circuit. Conversely, with a low input voltage (V1 < 0.65 V), the switch is open, and the transistor behaves as an open circuit. This perspective is applicable to both digital circuits and power electronic circuits.

Figure 3.5 shows an amplifier circuit with a MOSFET and the simulated transfer characteristics. Compared with the simulation results from the BJT amplifier shown in figure 3.4(b), the transition region moves toward the center, since the threshold voltage of this MOSFET is about 2 V. In addition, the slope of the transition region is less steep than that of the BJT amplifier, resulting in a lower voltage gain of about −21.2 V/V, which is shown in the inset of figure 3.5(b).

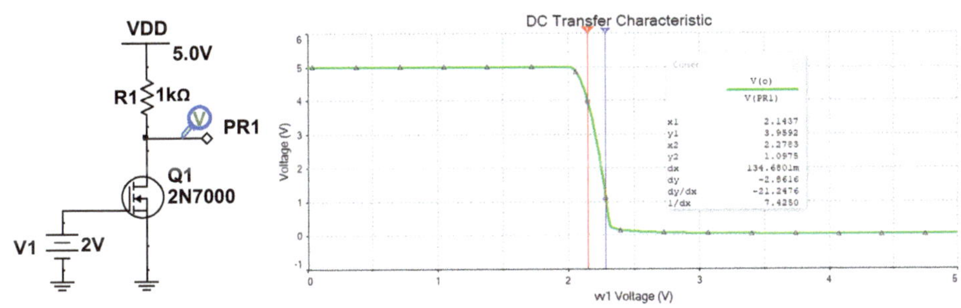

(a) (b)

Figure 3.5. Transfer characteristics of a MOSFET amplifier: (a) circuit, (b) simulation with DC sweep.

As introduced in chapter 2, the *I–V* characteristics of transistors can be categorized into three regions: cutoff, saturation/triode, and active. These regions directly correspond to the three distinct sections observed in the transfer characteristic curves.

- **Cutoff region:** the transistor current is negligible (open circuit), so the output voltage approaches the supply voltage (VCC or VDD).
- **Saturation/triode region:** the output voltage remains very low, but the transistor current is quite high.
- **Active region:** the transistor's behavior is similar to that of a voltage-controlled current source, which corresponds to the transition region in the transfer characteristic curve.

A key limitation of BJT transistors in digital logic applications is the delay experienced when exiting saturation mode. As covered in chapter 2, diffusion is the primary mechanism for minority carrier movement in BJTs, which is slower than the drift process. During saturation, minority carriers accumulate in the base region, causing a delay when switching the transistor off. In contrast, MOSFETs are unipolar devices that rely on the drift of majority carriers, eliminating this type of delay and making them generally preferred for digital logic circuits.

Figure 3.6 illustrates the switching behavior of BJT and MOSFET inverter circuits. The input signal displayed at the top is a square wave with a frequency of 500 kHz and a peak-to-peak voltage of 5 V. The output waveforms are depicted at the bottom, and a 0.1 nF load capacitor is connected to the output node. The output

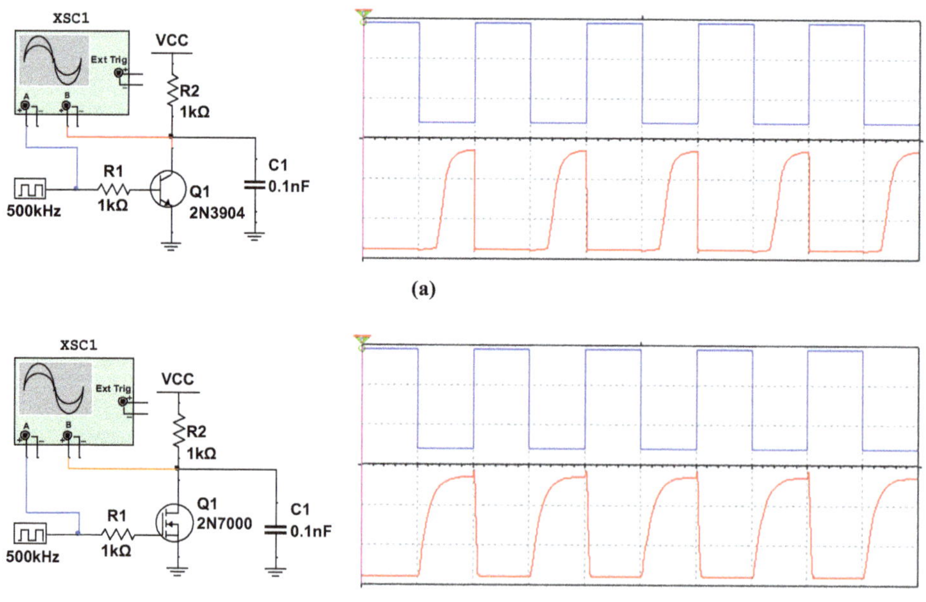

Figure 3.6. Delay in inverter circuits: (a) BJT inverter, (b) MOSFET inverter.

signal of the BJT inverter exhibits a noticeable delay in the rising edge, whereas the MOSFET inverter shows no such delay. In modern digital circuits operating at very high frequencies (~GHz), this delay in BJTs is unacceptable.

Additionally, both inverter circuits suffer from imbalanced pull-up and pull-down behavior. The pull-up curve, which relies on the current passing through a resistor, tends to be gradual, while the pull-down curve is typically sharper due to the higher current passing through the transistor. This imbalance can be addressed by replacing the pull-up resistor with a *p*-type transistor.

> Transistors offer a significant advantage over resistors in ICs, since they can provide much higher current at a smaller size. In fact, resistors occupy a considerably larger chip area compared to transistors, making them much more expensive than transistors. Consequently, designers strive to minimize the number of resistors in modern ICs, opting for transistors whenever possible.

3.3 DC operating conditions of amplifiers

The transfer characteristic curves shown in figures 3.4(b) and 3.5(b) demonstrate that the amplifier circuits only work in a very narrow voltage range at the input node. First, the transistors must be in active mode; otherwise, the amplifier does not work at all. Second, within the active mode, an optimized region is necessary to avoid output waveform distortion.

Figure 3.7 depicts the load line analysis curves, where the resistor acts as the load for this amplifier. The horizontal axis represents the collector-to-emitter voltage (V_{CE}), which is the same as V_C since the emitter is grounded. By applying Ohm's law to the resistor, its current can be found, namely $I_C = (V_{CC} - V_C)/R_1$. This relationship is represented by the straight line in the I–V diagram shown in figure 3.7(b), and the slope is $-1/R_1$. Conversely, each I–V curve corresponds to a specific base voltage (V_1) for the transistor. For example, if V_1 is low, it corresponds to a curve close to

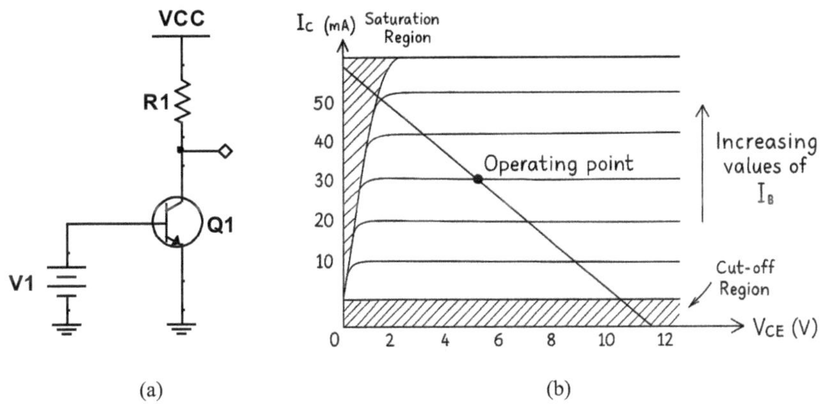

Figure 3.7. Load line analysis: (a) BJT amplifier circuit, (b) load line diagram. Created with GPT-4.0, OpenAI.

the cutoff region at the bottom. As V_1 increases, the curve shifts higher. Ideally, the operating point, which is also called the *Q-point*, should reside in the middle, as shown in the diagram, which enables the output signal to have a large amplitude without much distortion.

Figure 3.8(a) shows an amplifier circuit with a well-chosen DC bias point: $V_1 = 0.69$ V, $V_C = 2.44$ V. Notably, V_C sits halfway between V_{CC} and the emitter voltage, which is ground in this case. The peak-to-peak voltage of the output signal is 917 mV, so the resulting gain is $A_V = -91.7$ V/V. Figure 3.8(b) shows the waveforms, which exhibit good sine-wave shapes with little distortion.

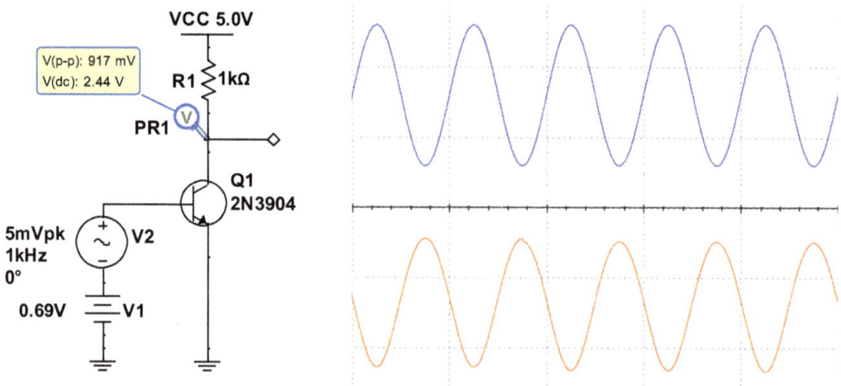

(a) (b)

Figure 3.8. Amplifier with a good DC bias point: (a) amplifier circuit, (b) waveforms.

Figure 3.9(a) illustrates the effect of increasing V_1 to 0.71 V. As a result, V_C drops to 0.267 V, and the transistor is close to saturation. In the load line diagram shown in figure 3.7(b), this behavior corresponds to a transistor curve at the top, and the intersection of the curve with the load line moves to the top left corner. Consequently,

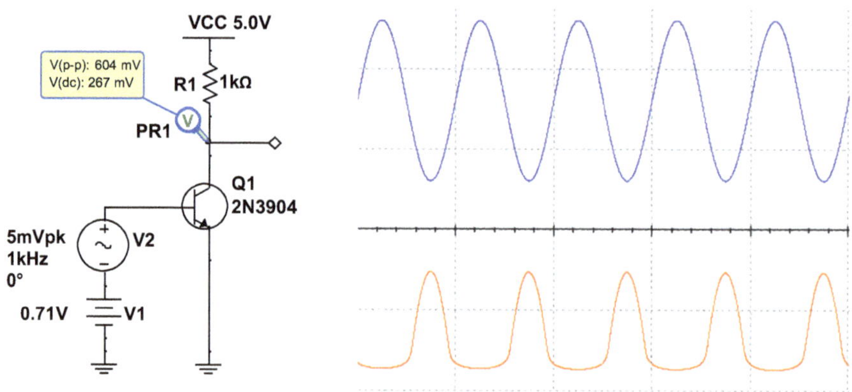

(a) (b)

Figure 3.9. Amplifier with a low DC collector voltage: (a) amplifier circuit, (b) waveforms.

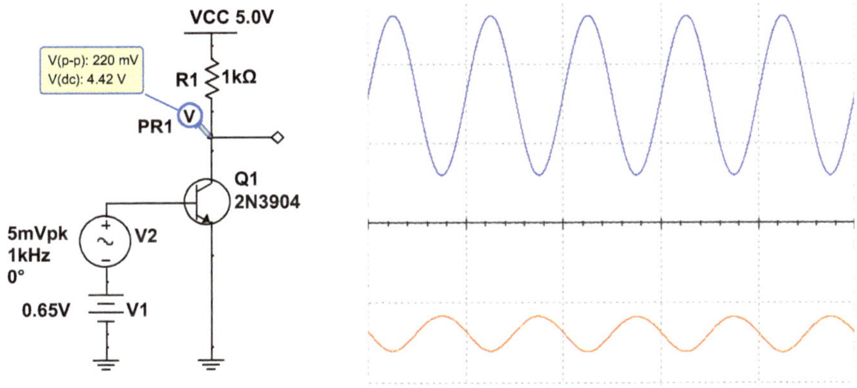

(a) (b)

Figure 3.10. Amplifier with a high DC collector voltage: (a) amplifier circuit, (b) waveforms.

the output waveform in figure 3.9(b) becomes severely distorted compared to a sine wave. Since V_C is already quite low, there is minimal room for further negative swings, limiting the amplifier's ability to faithfully amplify the input signal.

The opposite case is shown in figure 3.10(a): $V_1 = 0.65$ V. Here, $V_C = 4.42$ V, which is too close to V_{CC} and limits its ability to operate effectively. In addition, the simulation result indicates that the gain is reduced significantly: $A_V = -22$ V/V. In the load line diagram shown in figure 3.7(b), this situation corresponds to a transistor curve at the bottom, and its intersection with the load line shifts to the lower right corner. Unlike the waveform shown in figure 3.9(b), the waveform shown in figure 3.10(b) has minimal distortion, but its amplitude is significantly reduced. As discussed in chapter 2, the collector current (I_C) is an exponential function of the base–emitter voltage (V_{BE}), and the transconductance is directly related to I_C ($g_m = I_C/V_T$). Therefore, the lower collector current caused by the biasing condition results in a reduced gain.

In summary, the output node of amplifiers needs swing room for proper operation. Therefore, the DC bias point of this node should be at the midpoint between V_E and V_{CC} for the BJT amplifier circuits discussed in this section. Similarly, for MOSFET amplifier circuits with the same configuration, the DC bias point at the output node should be at the midpoint between V_S and V_{DD}. This is the objective of DC bias circuit design.

3.4 Design of DC biasing circuits

The bias circuits discussed in the previous section are very simple, but they are not practical, since there is usually no additional tunable DC voltage source available. A voltage divider circuit, like the one depicted in figure 3.11(a), offers a solution. If the

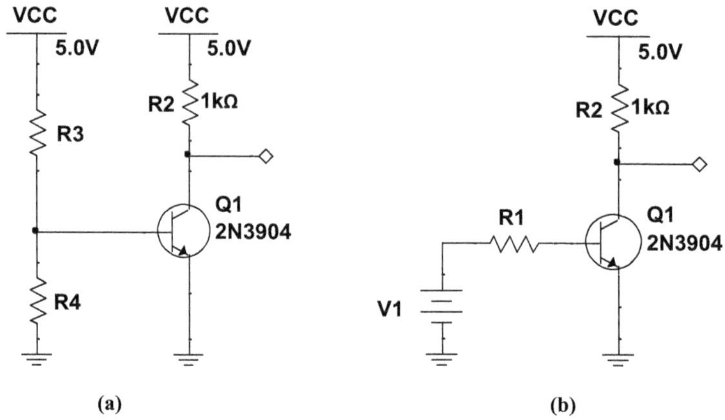

Figure 3.11. Simple amplifier bias circuit: (a) with voltage divider, (b) equivalent circuit.

transistor is replaced with a MOSFET, one can find the gate voltage directly using the voltage divider formula, since the gate current is zero. However, the base of a BJT needs some input current, so this is a *pseudo* voltage divider circuit, and its standard formula does not apply here. Thankfully, Thévenin's theorem allows us to transform this circuit into the equivalent one shown in figure 3.11(b). The following equations can be used to find the parameters of these equivalent circuit elements:

$$V_1 = \frac{R_4}{R_3 + R_4} V_{CC}, \quad R_1 = R_3 \,//\, R_4. \tag{3.4}$$

From the perspective of circuit design, the second circuit shown in figure 3.11(b) is easier to handle.

- Assuming the DC voltage at the collector is 2.5 V, then the collector current can be found from Ohm's law: $I_C = 2.5$ mA.
- For amplifier circuits, the BJT must be in active mode: $I_C = \beta I_B$. Therefore, the base current can be found: $I_B = I_C/\beta$.
- Assuming $V_{BE} = 0.7$ V, then $V_1 = 0.7 + I_B R_1$.

If $\beta = 250$, then $I_B = 10$ μA. Select $R_1 = 10$ kΩ, then $V1 = 0.8$ V. Using the two equations listed in equation (3.4), R_3 and R_4 can be determined: $R_3 = 62.5$ kΩ and $R_4 = 11.9$ kΩ. These values can be used as the initial design, and they need to be optimized through iterative circuit simulations.

The analysis presented above overlooked two crucial factors affecting transistor biasing. Firstly, the current gain (β) of the transistor is often unspecified. But more importantly, β itself varies significantly with temperature, as illustrated in figure 2.23 of the previous chapter. This variation can be substantial, especially during initial power-up processes when the transistor temperature rises quickly. To ensure consistent operation, we must design a bias circuit that minimizes its sensitivity to β fluctuations.

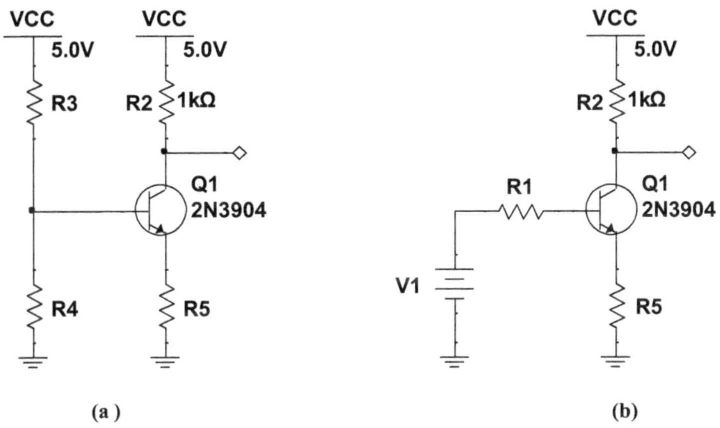

Figure 3.12. Stable amplifier bias circuit: (a) with voltage divider, (b) equivalent circuit.

In general, stability can be achieved with negative feedback. For the circuits shown in figure 3.11, negative feedback can be implemented by adding a resistor below the emitter of the transistor, as shown in figure 3.12. This resistor is called the 'emitter degeneration resistor.' Since I_C is tied to the collector voltage, it is required to be stable. Similarly, the emitter current (I_E) is very close to I_C ($I_C = \alpha I_E \approx I_E$), so it does not change either. On the other hand, the base current (I_B) depends on β directly.

Using Kirchhoff's voltage law (KVL), an equation can be set up for the loop at the bottom of the circuit in figure 3.12(b):

$$V_1 = I_B R_1 + 0.7 + I_E R_5. \tag{3.5}$$

Achieving high stability against β variations in equation (3.5) requires meeting this condition: $I_E R_5 \gg I_B R_1$. The first option is to use smaller resistors for R_3 and R_4, resulting in lower R_1, but this approach introduces other design challenges, which are discussed in more detail later. The second option is to use a larger resistor R_5, but this reduces the swing room of the output signal. In practice, R_5 is usually kept equal to or less than R_2 to balance these competing factors. For example, a conservative design partitions V_{CC} equally into three parts: $V_{R2} = V_{CE} = V_{R5}$. In a good design for a DC bias circuit, the collector voltage may exhibit slight shifts when β undergoes significant changes.

> If resistors R3 and R4 have lower resistances, a high current flows through this branch, leading to increased power consumption. Additionally, the amplifier's input resistance decreases, consequently reducing the voltage gain.

Multisim offers a powerful feature—sensitivity analysis—which helps visualize how design choices affect circuit stability. One simple approach is using a *parameter sweep analysis* to observe how changes in temperature or component values influence key circuit parameters, such as voltages or currents. To illustrate, a circuit without an emitter resistor is first analyzed, and the result is shown in figure 3.13. When the BJT junction temperature is varied from −20 °C to 40 °C, the collector voltage of the BJT shifts significantly, from 4.5 V down to 2.7 V, indicating high sensitivity to temperature changes. This instability underscores the importance of adding emitter degeneration, which can greatly reduce such variations by introducing negative feedback and thereby improving the robustness of the biasing scheme.

Figure 3.14 presents the results of the same sensitivity analysis, now applied to a circuit that includes an emitter degeneration resistor. As shown in figure 3.14(b), the collector voltage exhibits a much smaller shift—only about 0.3 V—over the same temperature range from −20 °C to 40 °C. This significant reduction in voltage drift clearly demonstrates the stabilizing effect of the emitter resistor. As a result, the biasing circuit becomes far less sensitive to temperature changes, greatly improving the reliability and predictability of amplifier performance.

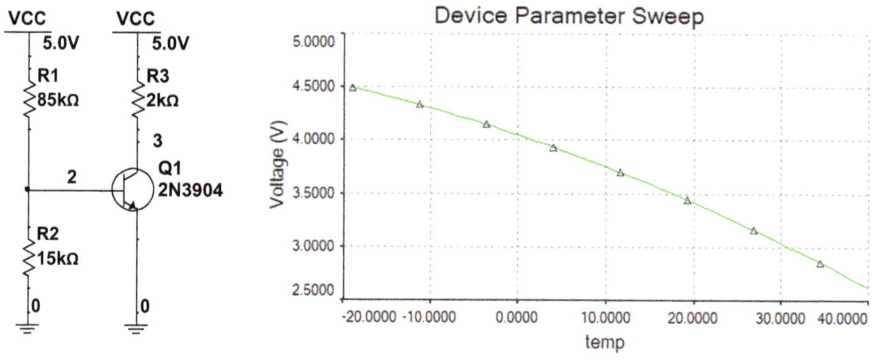

(a) **(b)**

Figure 3.13. Stability analysis: (a) circuit without feedback resistor, (b) drift with temperature.

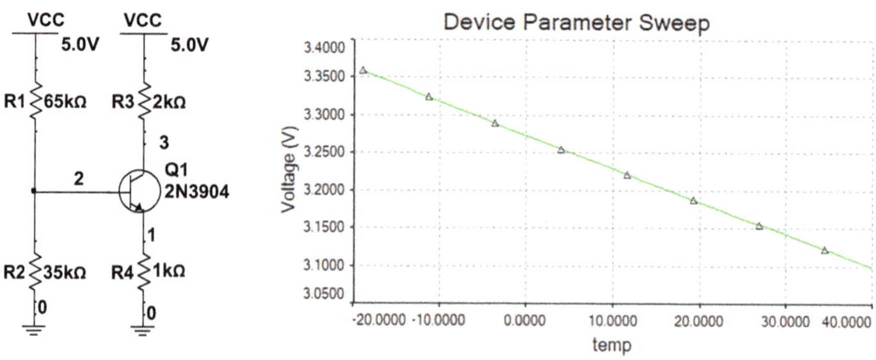

(a) **(b)**

Figure 3.14. Stability analysis: (a) circuit without feedback resistor, (b) drift with temperature.

	Variable	Sensitivity
1	qq1_area	-243.82354 m
2	qq1_temp	-37.22741 m
3	rr1	135.24449 u
4	rr2	-663.73948 u
5	rr3	-903.07697 u
6	vccvcc	-1.67416

(a)

	Variable	Sensitivity
1	qq1_area	-8.86117 m
2	qq1_temp	-4.32038 m
3	rr1	30.83547 u
4	rr2	-49.73225 u
5	rr3	-919.63248 u
6	vccvcc	407.92676 m

(b)

Figure 3.15. Results of sensitivity analysis: (a) without a feedback resistor, (b) with a feedback resistor.

In addition to 'parameter sweep analysis,' Multisim also offers a more sophisticated sensitivity analysis tool, which can show the sensitivity to several parameters. If the collector voltage is considered a function, then it depends on many parameters in the circuit, such as the transistor, resistors, and the DC voltage source. Therefore, variations in any of these parameters cause changes in the collector voltage, and this relationship can be expressed in the following equation:

$$\mathrm{d}V = \frac{\partial V}{\partial A}\mathrm{d}A + \frac{\partial V}{\partial T}\mathrm{d}T + \sum_k \frac{\partial V}{\partial R_\mathrm{k}}\mathrm{d}R_\mathrm{k} + \frac{\partial V}{\partial V_\mathrm{CC}}\mathrm{d}V_\mathrm{CC}. \tag{3.6}$$

The first two terms in the equation above correspond to changes in transistor size and temperature. The simulation results shown in figure 3.15 are the derivatives in equation (3.6), and thus a lower value indicates lower sensitivity and higher stability. The data reveal the dramatic improvement in stability achieved with the emitter degeneration resistor. Sensitivity to all parameters except R_3 is significantly reduced. For instance, sensitivity to transistor size shrinks to a mere 3.63% compared to the original circuit. Similarly, sensitivity to transistor temperature plummets by nearly an order of magnitude. These remarkable results underscore the critical role of the emitter degeneration resistor in ensuring robust performance for practical amplifier circuits.

3.5 Transistor models

A bipolar junction transistor (BJT) has two *pn* junctions, as shown in figure 3.16(a). However, their roles differ significantly. When a BJT operates in the forward-active mode, the base–emitter junction is forward biased and behaves like a regular diode, making the current very sensitive to the voltage across this junction (V_BE). In contrast, the current has a very weak dependence on the voltage across the base–collector junction (V_CB) or the collector–emitter voltage (V_CE), as reflected in the BJT's characteristic curves. From the perspective of the collector node, the BJT acts like a current source controlled mainly by the base–emitter voltage. Based on these features, a circuit model can be created, as shown in figure 3.16(b). Due to its resemblance to the letter 'T,' this model is called the T-model. It is important to

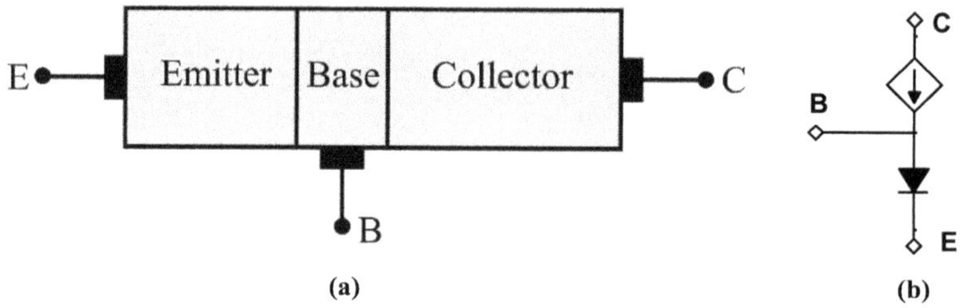

Figure 3.16. Device structure and circuit model: (a) BJT structure, (b) large-signal T-model. Created with GPT-4.0, OpenAI.

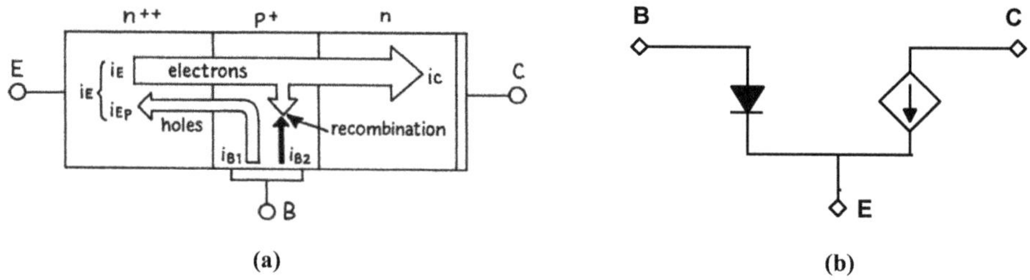

Figure 3.17. Currents and circuit model: (a) currents in a BJT, (b) large-signal hybrid-π model. Created with GPT-4.0, OpenAI.

emphasize that this model only works when the transistor is in the forward-active mode.

The relationship between the three terminal currents is governed by Kirchhoff's current law (KCL): $I_B + I_C = I_E$. In active mode, a simple relationship exists between the collector and base currents: $I_C = \beta I_B$. A similar relationship can also be specified between the collector and emitter currents: $I_C = \alpha I_E$, where $\alpha = \beta/(\beta + 1)$. This formula can be derived from KCL, and it is called 'common-base current gain.' In general, $\beta > 100$, $\alpha \approx 1$, so the base current is just like a small creek joining a large river at the intersection.

Figure 3.17(a) depicts various currents inside a BJT, and it shows that the collector current is flowing across the base directly between the emitter and collector. Therefore, a new circuit model can be constructed, which is shown in figure 3.17(b). Due to its resemblance to the inverted Greek letter Π, it is called the hybrid-π model. A key distinction between this model and the T-model lies in how the diode current is represented. In the hybrid-π model, the diode current is the base current, which is significantly smaller than the diode current (emitter current) in the T-model.

The circuit models shown in figures 3.16(b) and 3.17(b) are oversimplified because they ignore the Early effect. This assumption implies that the I–V curves in figure 3.1(b) are perfectly flat. In reality, there is a small slope in these curves. This slope can be modeled by adding a resistor between the collector and emitter nodes in these circuit models. While incorporating this adjustment into the hybrid-π model poses no significant challenges for circuit analysis, the revised T-model becomes overly complicated and impractical. Therefore, the hybrid-π model is generally preferred when

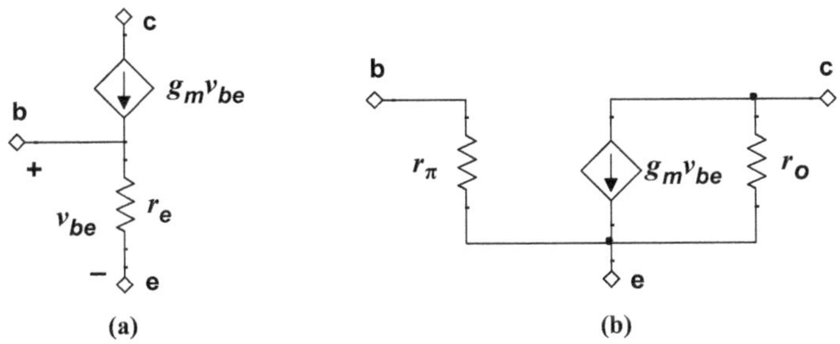

Figure 3.18. Low-frequency small-signal models of a BJT: (a) T-model, (b) hybrid-π model.

considering the Early effect. For example, in ICs, the channel length of MOSFETs is very short, making the Early effect significant and necessitating the use of the hybrid-π model.

In amplifier circuits, the voltages and currents have both DC and AC components. Typically, the input AC signal is very weak, allowing the circuit models discussed above to be further simplified into the so-called 'small-signal models.' In the low-frequency domain, a diode can be replaced by its incremental resistance. This transformation is applied to both the T-model and the hybrid-π model, resulting in the small-signal models depicted in figure 3.18.

The parameters for the BJT transistors in these models are specified below; they are related to the DC terminal currents.

- Transconductance (g_m): is determined by the collector current: $g_m = I_C/V_T$, where $V_T = 25.9$ mV at room temperature.
- Incremental resistance (r_e): the T-model features an emitter resistance inversely proportional to the emitter current (I_E): $r_e = V_T/I_E$. This resistance is relatively low; for an emitter current of 1 mA, $r_e \approx 25.9\ \Omega$.
- Incremental resistance (r_π): this resistance in the hybrid-π model is inversely proportional to the base current (I_B): $r_\pi = V_T/I_B$. This resistance is much higher than r_e. For a transistor with a base current of 10 μA, $r_\pi \approx 2.59\ k\Omega$.
- Output resistance (r_o): the Early effect gives rise to the output resistance, which depends on the Early voltage (V_A) and collector current (I_C): $r_o = V_A/I_C$. Discrete transistors typically have high Early voltages. For example, if $V_A = 100$ V, $I_C = 1$ mA, then $r_o = 100$ kΩ. Therefore, the simplified T-model shown in figure 3.18(a) is a good approximation for most circuits with discrete components.

In small-signal models, these device parameters are determined by the DC parameters of the transistor. Therefore, it is imperative to analyze the DC bias circuit first before analyzing the AC circuit. Table 3.1 summarizes the formulae for the parameters in the small-signal models.

Similar low-frequency small-signal models exist for MOSFETs, as shown in figure 3.19. While these models share some resemblance to their BJT counterparts, key differences exist. Due to the gate's insulating oxide layer, there is no direct resistance between the gate and source in the hybrid-π model, as shown in figure 3.19(b).

Table 3.1. Formulae for small-signal model parameters.

Parameters	Formula
g_m	$g_m = I_C/V_T$
r_π	$r_\pi = V_T/I_B$
r_e	$r_e = V_T/I_E$
r_o	$r_o = V_A/I_C$
$g_m \sim r_\pi$	$g_m\ r_\pi = \beta$
$g_m \sim r_e$	$g_m\ r_e = \alpha$

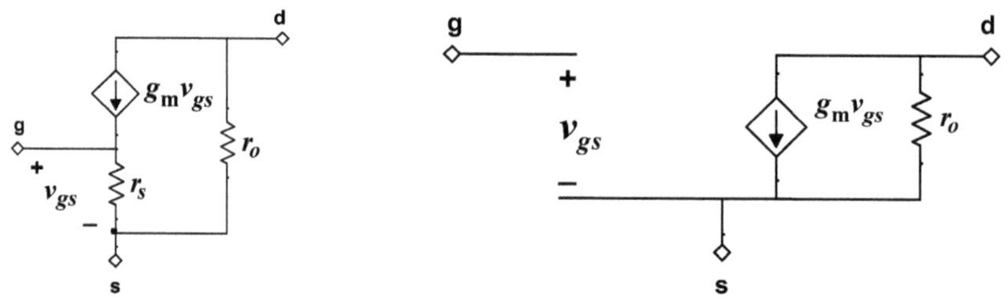

Figure 3.19. Low-frequency small-signal models of a MOSFET: (a) T-model, (b) hybrid-π model.

However, if one needs to analyze a MOSFET using a formula derived for a BJT, $r_\pi \to \infty$ can be assumed. In addition, the formula for the transconductance of MOSFETs is different; specifically, the transconductance is proportional to the square root of the drain current: $g_m = \sqrt{2kI_D}$. It also depends on a device parameter (k). For comparison, transconductance in BJTs is proportional to the collector current and does not depend on any device parameters.

In the T-model shown in figure 3.19(a), the gate appears to be directly connected to the conducting channel. However, to ensure zero gate current, the resistor r_s needs to satisfy this condition: $i_s = i_d = g_m v_{gs}$, which gives rise to the formula for this resistance: $r_s = 1/g_m$. Since MOSFETs are primarily used in ICs, the Early voltage (V_A) is typically much lower due to the 'short-channel' effect. Consequently, the output resistance (r_o) cannot be ignored, so it is included in the T-model in figure 3.19(a).

These circuit models shown in figure 3.19 are only valid for very weak input signals, which is why they are called 'small-signal models.' Mathematically, the parameters used (such as g_m and r_e) relate to the first-order derivatives of the variables. When the input signal becomes stronger, higher-order terms become significant. This introduces nonlinear behavior into the output waveform. Power amplifiers, which deal with strong input signals, require entirely different analysis methods.

3.6 CE/CS amplifier circuits

In single-transistor amplifiers, one of the three transistor terminals is typically connected to AC ground. In a circuit, the ground is a single and *common* reference point, though several ground symbols may appear in multiple locations. As a result, these amplifier configurations are categorized based on which terminal is AC grounded, and the names of these configurations are listed in table 3.2.

The CE/CS amplifier is the most useful configuration, since it has a balanced performance in most respects. Figure 3.20(a) shows an example of a CE amplifier, where all capacitors are assumed to be large. In DC analysis, these capacitors block the flow of DC current, so only the central part of the circuit remains, which was analyzed in section 3.4. In AC analysis, these large coupling capacitors act as short circuits, and the node at V_{CC} becomes AC grounded. This simplification allows us to analyze the AC equivalent circuit shown in figure 3.20(b).

Table 3.3 summarizes how certain circuit elements behave differently in DC and AC analyses. The rules for capacitors and inductors are related to their impedances: $|Z_C| = 1/\omega C$, $|Z_L| = \omega L$. Therefore, the criteria for large and small capacitors/inductors are frequency related. For example, the internal capacitances of transistors can be considered small and ignored at low and medium frequencies, but they play important roles at high frequencies, which are discussed in detail in section 3.8.

Tabel 3.2. Classification of amplifier circuits.

BJT	MOSFET
CE amplifier	CS amplifier
CB amplifier	CG amplifier
CC amplifier	CD amplifier

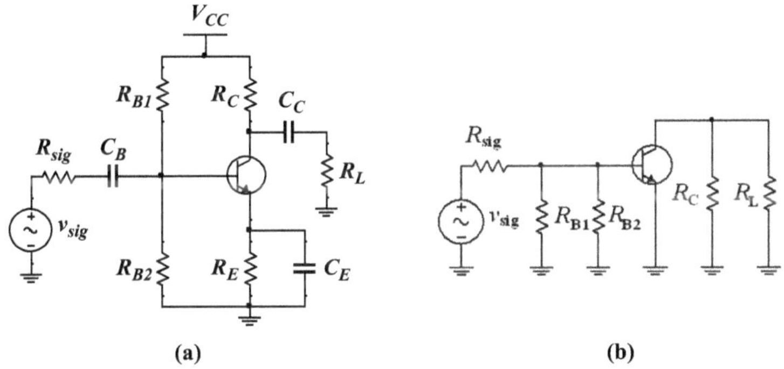

Figure 3.20. CE BJT amplifier circuit: (a) complete circuit, (b) equivalent AC circuit.

Table 3.3. Behaviors of circuit elements in DC and AC equivalent circuits.

Device	DC	AC
Resistor	Same	Same
Small capacitor	Open	Open
Large capacitor	Open	Short
Small inductor	Short	Short
Large inductor	Short	Open
DC V-source	Same	Short
AC V-source	Short	Same
DC I-source	Same	Open
AC I-source	Open	Same

- **Capacitors:** the impedances of capacitors reach infinity for DC ($\omega = 0$), so they can effectively block DC current and act like open circuits. In contrast, the impedances of large capacitors become very low for AC, essentially functioning as short circuits.
- **Inductors:** unlike capacitors, inductors exhibit very low impedance for DC. This allows direct current to flow easily through them, so they can be treated as short circuits except for their parasitic resistance. In AC analysis, however, the impedance of large inductors becomes very high. At high enough frequencies, this high impedance can be so significant that inductors can be treated as open circuits for AC signals.
- **Voltage sources:** voltage sources act as short circuits for their opposite types (AC vs DC) because that voltage component becomes zero. For example, a DC voltage source acts as a short circuit in the equivalent AC circuit. Therefore, V_{CC} and V_{DD} become ground in AC analysis.
- **Current sources:** current sources act as open circuits for their opposite types (AC vs DC) because that current component becomes zero. For example, a DC current source acts as an open circuit in the equivalent AC circuit.

For ICs, inductors are mainly used in RF/microwave circuits, which operate at very high frequencies. Inductors occupy a very large chip area, so only very small inductors can be fabricated in ICs. As we know, the impedance of an inductor is proportional to its inductance and the frequency: $|Z_L| = \omega L$. Therefore, small inductors can induce a large impedance at high frequencies, so they are very effective in RF/microwave circuits.

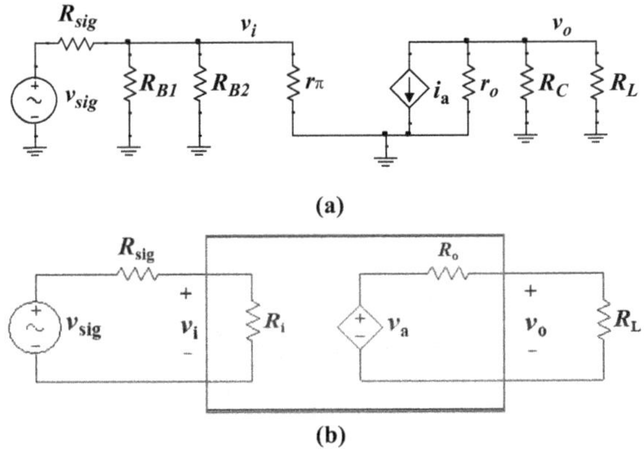

Figure 3.21. Small-signal circuit of a CE BJT amplifier: (a) original circuit, (b) equivalent circuit.

In the next step, the BJT in figure 3.20(b) is replaced with its small-signal model, and the resultant circuit is shown in figure 3.21(a). This circuit can be simplified into figure 3.21(b) via the following transformations:

- **Combining resistors:**
- On the left-hand side, resistors R_{D1}, R_{D2}, and r_π are in parallel. We can combine them into a single 'input resistance': $R_i = R_{B1} // R_{B2} // r_\pi$.
- Similarly, on the right-hand side, resistors R_C and r_o are in parallel. Their combination forms the 'output resistance': $R_o = R_C // r_o$.
- **Thévenin's equivalent:**

  The current source in parallel with the output resistance (R_o) can be converted into its Thévenin equivalent circuit. The resultant value of the voltage source is $v_a = -i_a R_o$, where the negative sign comes from the direction of the current source. Since $i_a = g_m v_i$, the value of the voltage source can be found: $v_a = -(g_m R_o) v_i$.

The coupling capacitors C_B and C_C in figure 3.20(a) define the amplifier circuit's boundaries. Consequently, the simplified circuit shown in figure 3.21(b) highlights this separation between the core amplifier (inside the block) and its peripherals: the signal source on the left and the load on the right. In addition, an amplifier's core functionality is often characterized by three key parameters: voltage gain (A_{vo}), input resistance (R_i), and output resistance (R_o). These parameters are mathematically represented in the following equations:

$$R_i = R_{B1} \| R_{B2} \| r_\pi, \quad R_o = r_o \| R_C, \quad A_{VO} = \frac{v_a}{v_i} = -g_m \cdot R_O. \tag{3.7}$$

The voltage gain of this amplifier circuit can be readily analyzed from the simplified circuit in figure 3.21(b), which can be visualized as a cascade of three stages: the core amplifier and a voltage divider on each side. By considering the voltage changes across each stage, the overall gain can be derived as the product of the individual gains of these three sections:

$$A_V = \frac{v_o}{v_{sig}} = \frac{v_i}{v_{sig}}\frac{v_a}{v_i}\frac{v_o}{v_a} = -\frac{R_i}{R_{sig}+R_i}g_m R_o \frac{R_L}{R_o+R_L}. \quad (3.8)$$

Understanding the distinction between the amplifier gain (A_V) and the core amplifier gain (A_{VO}) can be aided by an analogy: imagine the core amplifier gain (A_{VO}) as your gross salary; the amplifier gain (A_V) is like your 'purchasing power' after taxes and other deductions. First, one cannot receive the full salary, since a few taxes are deducted first, which is equivalent to the first factor in equation (3.8). Second, when one buys something, the sales tax needs to be paid, which is equivalent to the third factor in equation (3.8). Based on this equation, three key characteristics are desirable for the amplifier to achieve high overall gain:

- **High transconductance (g_m)**: this parameter signifies the amplifier's capability to convert a voltage signal into a current signal. A higher g_m translates to a higher voltage gain for the core amplifier ($g_m R_o$).
- **High input impedance (R_i)**: a high input impedance minimizes the current drawn by the amplifier from the signal source, preventing signal attenuation. This is crucial for maintaining a strong input signal to the transistor.
- **Low output impedance (R_o)**: a low output impedance reduces the voltage drop across the amplifier's output resistance when delivering current to the load. This ensures that the amplified signal is efficiently transferred to the load.

> The criteria discussed above apply specifically to voltage amplifiers, the most common type. However, amplifiers come in various categories depending on the type of signals they handle. Here are three additional amplifier types based on input and output signals: (a) current amplifier, (b) transimpedance amplifier (current input, voltage output), (c) transadmittance amplifier (voltage input, current output). The requirements for these amplifier types differ from those of voltage amplifiers.

Figure 3.22 shows an example of a CE amplifier circuit. The simulation results show that the magnitude of the overall voltage gain is 17.9 V/V. In addition, from the DC voltage at the collector, we can find the collector current: $I_C = 1$ mA, and then $g_m \approx$ 38.6 mS. The current gain can be found from the simulation: $\beta = 135$, and then the base current can be calculated: $I_B \approx 7.42$ μA; therefore, $r_\pi \approx 3.49$ kΩ. The Early voltage can be found from the transistor model, $V_A = 74$ V, and then $r_o = 74$ kΩ.

We can now calculate the three key parameters of the core amplifier:

- **Input impedance:** $R_i = R_{B1}//R_{B2}//r_\pi \approx 2.28$ kΩ
- **Output impedance:** $R_o = r_o//R_C \approx 1.95$ kΩ
- **Core voltage gain:** $A_{VO} = -g_m \cdot R_O = -75.3$ V/V.

The first factor in equation (3.8) (income-related taxes) is 0.695, and this value is slightly less than the simulation result (0.715). The third factor in equation (3.8) (sales tax) is 0.339, which contributes the most significant decrease in gain.

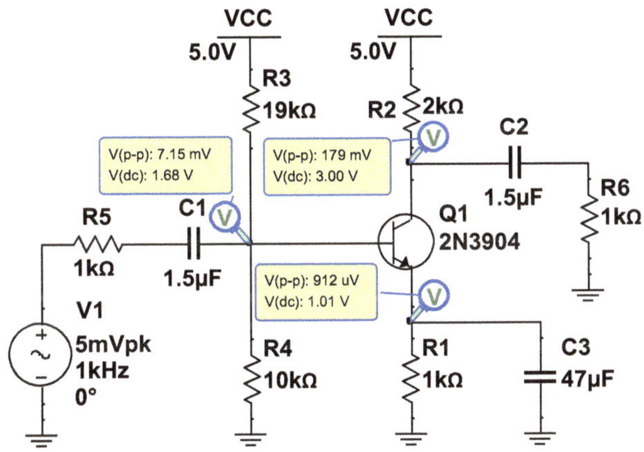

Figure 3.22. An example of a CE BJT amplifier.

By multiplying these factors together with the core gain (A_{VO}), we obtain the overall voltage gain (A_V) of approximately −17.7 V/V. This closely matches the simulation result of −17.9 V/V, validating the analysis.

Amplifier designers have limited control over peripheral elements such as the signal source and load. However, they can optimize the amplifier circuit itself. This includes selecting appropriate values for the four resistors (R1–R4) within the core amplifier and the three capacitors (C1–C3). Building on the DC bias circuit discussed in section 3.4, we can refine the design presented in figure 3.22 to achieve higher gain. To accomplish this, we can decrease the value of R2 so that the output resistance is lower, but the other three resistors would then need adjustments to maintain the appropriate DC bias point.

It is important to remember that the DC and AC characteristics are interrelated, so one cannot over-optimize a specific parameter. For instance, lowering R_3 and R_4 can reduce temperature sensitivity in the DC bias circuit, but this comes at the cost of reduced input impedance and lower overall gain. Additionally, decreasing these two resistor values also increases the power consumption in this path, which is undesirable for battery-powered portable electronic devices.

In addition to the resistors, capacitor selection is also very important. If the frequency of the signal source is fixed, the design process becomes relatively straightforward. The next section explores the impact of these capacitors on low-frequency response.

- **Capacitor design criterion:** the primary goal for capacitor selection is to ensure that the capacitors act as short circuits at the desired frequency. This effectively removes them from the AC circuit during analysis. The criterion is $|Z_C| \ll R$, where R stands for a reference resistor, which is usually the resistor in series with the capacitor.
- **Coupling capacitors (C1 & C2):** designing C1 and C2 is relatively straightforward. The resistors R5 and R6 in figure 3.22 are chosen as the reference

resistors because the amplifier's input and output resistances are expected to be higher than these resistor values in this circuit.

- **Bypass capacitor (C3):** designing this capacitor presents a bigger challenge. Here, the reference resistor is the emitter resistance (r_e) in the T-model, which is much lower than R1. To compensate for this low resistance, a larger capacitor in figure 3.22 is needed. The simulation results indicate that the emitter node is effectively grounded for AC, since the signal level there is attenuated to a negligible level (less than 1 mV).

Just like the CE BJT amplifier, the CS MOSFET amplifier is the most used configuration in discrete circuits. Figure 3.23(a) shows an example of a CS amplifier circuit, where all the coupling and bypass capacitors are assumed to be large. In DC analysis, the voltage divider section (R_1 and R_2) is simpler to analyze compared to the BJT circuit. Because the gate is insulated, the DC voltage at the gate can be readily calculated using the standard voltage divider formula: $V_G = [R_2/(R_1 + R_2)]V_{DD}$.

In AC analysis, these capacitors become short circuits, and the node at V_{DD} is grounded. In addition, the MOSFET is replaced with its small-signal model, and the resulting circuit is shown in figure 3.23(b). Similar to the CE amplifier, this CS amplifier circuit can be conceptually divided into three sections: the signal source, the core amplifier, and the load, as illustrated in figure 3.23(c). However, the current

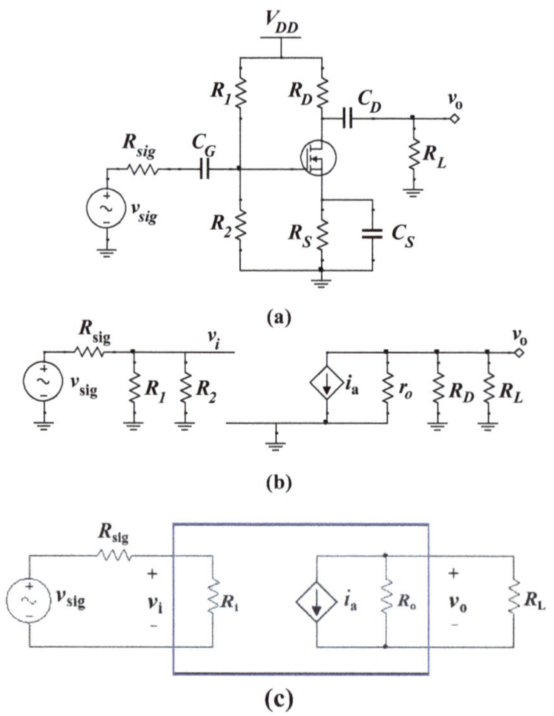

Figure 3.23. CS amplifier: (a) complete circuit, (b) small-signal circuit, (c) simplified circuit.

source is not converted into a voltage source and remains in Norton's format instead. The input and output resistances of the core amplifier can be found using the following equations:

$$R_i = R_1 || R_2, \quad R_o = r_o || R_D. \tag{3.9}$$

One key advantage of MOSFETs is their insulated gate. This allows the biasing resistors (R_1 and R_2) to have very high values, typically in the megohm (MΩ) range. As a result, the amplifier's input resistance becomes very high. This high input resistance minimizes the current drawn from the signal source, ensuring a strong input signal. Mathematically, the signal at the gate is very close to the signal from the source: $v_i \approx v_{sig}$. In addition, we can derive the voltage gain of the amplifier directly without converting the current source to a voltage source. From this perspective, the load resistor (R_L) is in parallel with the amplifier's output resistance (R_o):

$$i_a = g_m v_i \approx g_m v_{sig}, \quad v_o = -i_a \cdot (R_O || R_L), \quad A_V = \frac{v_o}{v_{sig}} \approx -g_m \cdot (R_O || R_L). \tag{3.10}$$

This result bears a resemblance to the gain expression obtained for the CE BJT amplifier, so the gain of the CS amplifier can be represented using a comparable mathematical format:

$$A_V = -\frac{R_i}{R_{sig} + R_i} g_m R_O \frac{R_L}{R_O + R_L} \approx -g_m \cdot (R_O || R_L). \tag{3.11}$$

> MOSFETs are widely used in ICs due to their favorable characteristics. However, for discrete small-signal amplifiers, BJTs are superior to MOSFETs, since their transconductance is higher. In addition, Multisim offers good device models for many BJTs, but few models for MOSFETs. Furthermore, the MOSFET models offered are very primitive. Therefore, BJT circuits are used for most simulations in this book.

3.7 Low-frequency response of CE/CS amplifiers

Multisim's Bode plotter is a valuable tool for visualizing an amplifier's frequency response, as shown in figure 3.24 for a CE amplifier. The plot shows the constant gain behavior that occurs within the midband region. For AC analysis in the midband, we can simplify the circuit by treating the coupling and bypass capacitors as short circuits and the internal transistor capacitors as open circuits. This section focuses on the low-frequency response influenced by the external capacitors. The high-frequency behavior due to the internal transistor capacitances is discussed in the next section.

The impact of coupling capacitors C1 and C2 can be readily analyzed using the small-signal circuit from figure 3.21(b). With these capacitors inserted, we obtain

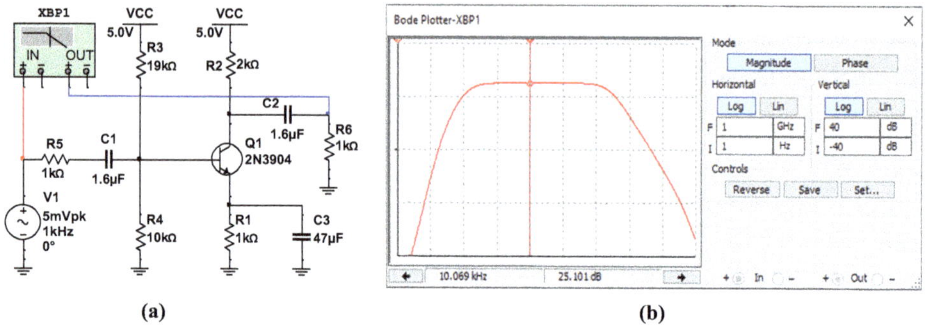

Figure 3.24. Frequency response of a CE amplifier: (a) amplifier circuit, (b) Bode plot.

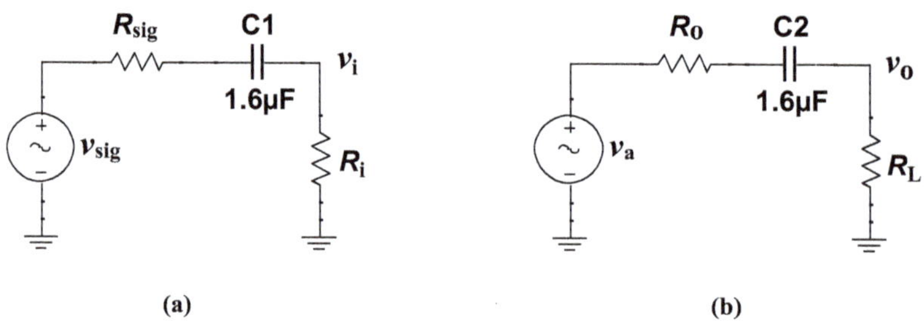

Figure 3.25. Equivalent circuit with coupling capacitors: (a) input circuit with C1, (b) output circuit with C2.

two similar circuits, as shown in figure 3.25. As discussed in chapter 1, this type of circuit behaves as a high-pass filter with the magnitude attenuated in the pass band. In circuit analysis, the capacitor can first be swapped with the resistor on its left-hand side; then, it becomes a standard high-pass filter with a voltage divider. Therefore, the transfer function is the product of these two subcircuits:

$$H_{C1}(s) = \frac{R_i}{R_{sig} + R_i}\frac{s}{s + \omega_{C1}}, \quad H_{C2}(s) = \frac{R_L}{R_o + R_L}\frac{s}{s + \omega_{C2}}, \tag{3.12}$$

where $\omega_{C1} = 1/[(R_{sig} + R_i)C_1]$ and $\omega_{C2} = 1/[(R_o + R_L)C_2]$. Using the parameters shown in figure 3.24(a), these two cutoff frequencies can be calculated: $f_{C1} \approx 30.3$ Hz, $f_{C2} \approx 33.7$ Hz.

The analysis of the bypass capacitor C3 in figure 3.24(a) is a little challenging. First, the T-model of the transistor can be used, which results in the subcircuit shown in figure 3.26(a). The input signal is at the base node (v_i), and the output signal is taken across the base–emitter junction (v_{be}). Next, this circuit can be transformed while keeping the output signal across r_e on the right, as shown in figure 3.26(b). In chapter 1, we analyzed this circuit (figure 1.18), called a 'lead compensator circuit,' and derived its transfer function:

$$H_{C3}(s) = \frac{s + \omega_z}{s + \omega_{C3}}, \tag{3.13}$$

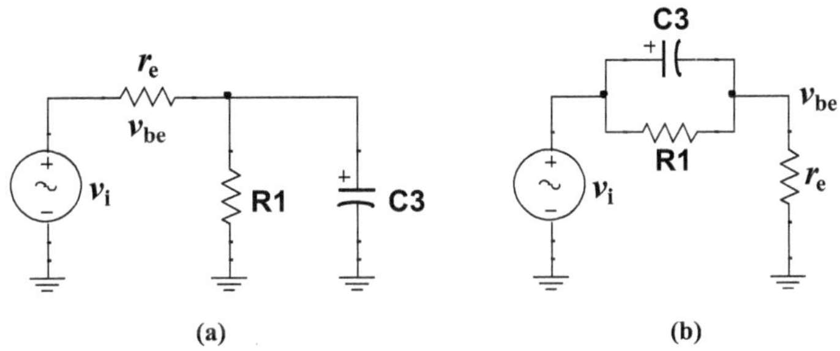

Figure 3.26. Equivalent circuit with bypass capacitors: (a) original circuit, (b) transformed circuit.

where $\omega_z = 1/(R_1 C_3)$ and $\omega_{C3} = 1/[(R_1 \| r_e)C_3] \approx 1/(r_e C_3)$. Using the parameters shown in figure 3.24(a), these two cutoff frequencies can be calculated: $f_z \approx 3.4$ Hz, $f_{C3} \approx 131$ Hz.

When the transfer functions of the three individual capacitors are taken into account, the voltage gain of the amplifier is modified as follows:

$$A_V(s) = \frac{v_o}{v_{sig}} = -\left(\frac{R_i}{R_{sig} + R_i} g_m R_o \frac{R_L}{R_o + R_L}\right)\left(\frac{s}{s + \omega_{C1}} \cdot \frac{s}{s + \omega_{C2}} \cdot \frac{s + \omega_z}{s + \omega_{C3}}\right). \quad (3.14)$$

Since these three poles are close to one another, there is no dominant pole. In this case, all of them contribute to the cutoff frequency:

$$f_L = \sqrt{f_{C1}^2 + f_{C2}^2 + f_{C3}^2 - 2f_z^2}. \quad (3.15)$$

For the circuit shown in figure 3.24(a), the cutoff frequency on the low-frequency side is $f_L \approx 139$ Hz, and this agrees quite well with the simulation results. This cutoff frequency is primarily determined by the bypass capacitor C3. However, the effects of C1 and C2 have moved it a little higher. Therefore, increasing the value of C3 would effectively lower the cutoff frequency, up to the point where the roles of C1 and C2 become dominant.

> The zero in the low-frequency domain effectively cancels a pole, so the slope of the Bode plot on the low-frequency side is 40 dB/dec instead of 60 dB/dec for three poles. This conclusion can be verified from the simulated results shown in figure 3.24(b).

3.8 High-frequency response of CE/CS amplifiers

The Bode plot in figure 3.24(b) shows that the gain rolls off in both the low- and high-frequency domains. The primary factors responsible for the high-frequency roll-off are the internal transistor capacitances, as illustrated in figure 3.27. Bandwidth is a crucial characteristic of amplifiers. It is defined by the high-frequency cutoff frequency, where the gain decreases by 3 dB from its midband value.

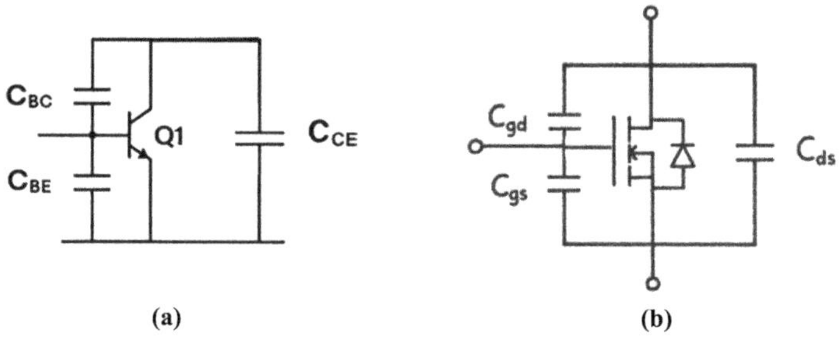

Figure 3.27. Internal capacitances of transistors: (a) BJT, (b) MOSFET. Created with GPT-4.0, OpenAI.

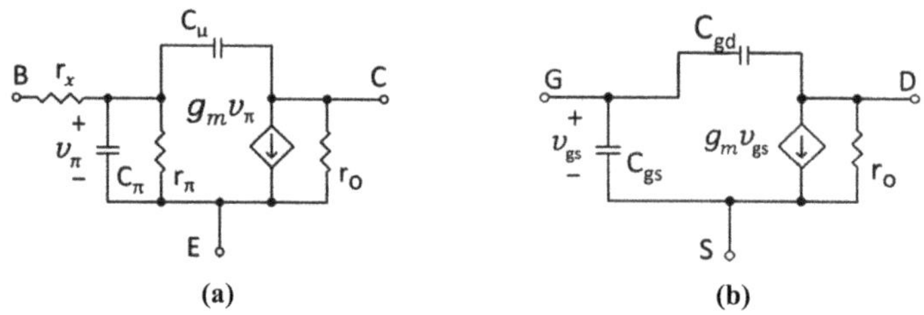

Figure 3.28. High-frequency hybrid-π models: (a) BJT, (b) MOSFET. Created with GPT-4.0, OpenAI.

Among these three internal capacitors, C_{CE} for BJTs and C_{DS} for MOSFETs are the least significant and are often negligible in many scenarios. Conversely, the largest capacitance is C_{BE} for BJTs or C_{GS} for MOSFETs in the forward-active mode. However, for CE/CS amplifiers, the capacitors $C_{\mathrm{BC}}/C_{\mathrm{GD}}$ hold greater importance, since they are effectively amplified. Consequently, CE/CS amplifiers exhibit the lowest bandwidth compared to alternative amplifier configurations.

With C_{CE} or C_{DS} disregarded, the remaining two capacitors can be integrated into the hybrid-π model, as depicted in figure 3.28. In the BJT model, these are conventionally referred to as C_π and C_μ, respectively. The presence of the capacitor between the input and output nodes complicates circuit analysis, prompting the widespread application of Miller's theorem to split this capacitor into two on either side.

Miller's theorem is a powerful tool used in analyzing the behavior of amplifiers at high frequencies. It deals with the situation where an impedance is present between the input and output terminals of an amplifier, such as the inherent capacitances within transistors. Figure 3.29 shows the equivalent circuit for an amplifier with a feedback capacitor, provided the gain of the amplifier is negative ($-K$). The formulae for these two equivalent capacitors can be derived as follows:

$$C_{\mathrm{M1}} = (1 + K)C_{\mathrm{f}}, \quad C_{\mathrm{M2}} = (1 + 1/K)C_{\mathrm{f}}. \tag{3.16}$$

When the magnitude of the gain K is high, C_{M1} is much higher than C_{f}, but C_{M2} is very close to C_{f}.

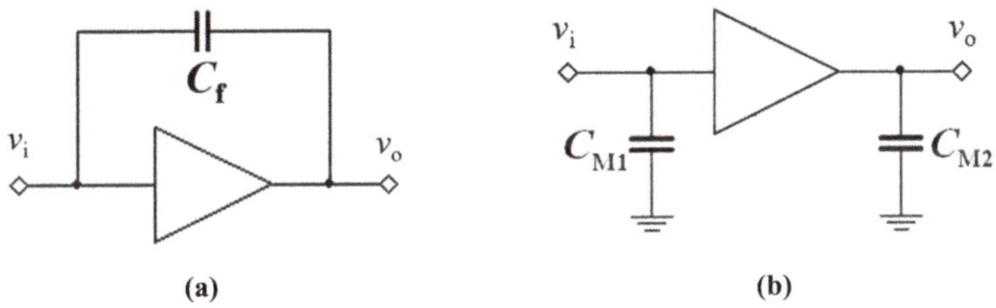

Figure 3.29. Miller's theorem: (a) amplifier with feedback capacitor, (b) equivalent circuit. Created with GPT-4.0, OpenAI.

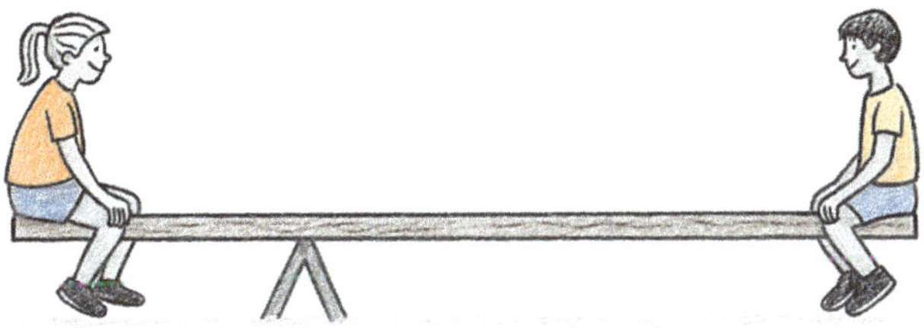

Figure 3.30. Intuitive analysis of Miller's theorem.

While Miller's theorem has a rigorous mathematical derivation, an intuitive explanation using an analogy can be helpful. Imagine an asymmetric seesaw like the one shown in figure 3.30. Assume $K = 9$, such that 10% of the beam is on the left-hand side and 90% of it is on the right-hand side. In this analogy, the input signal has the effect of moving the left end up and down, and the output signal on the right end moves accordingly. However, the fulcrum point remains stationary, which acts like the AC ground in a circuit. Since the fulcrum does not move, we can conceptually cut the seesaw at this point. The resulting two sections can be viewed as equivalent to the two impedances of the Miller capacitors, C_{M1} and C_{M2}, as shown in figure 3.29(b).

Using this intuitive model, two equations can be set up:

- The length of the entire beam (Z_f) is the sum of the two sections: $Z_{M1} + Z_{M2} = Z_f$.
- The ratio of these two sections is equal to the voltage gain: $Z_{M2} = K \cdot Z_{M1}$.

The solutions of these two equations can be found easily:

$$Z_{M1} = \frac{1}{1+K} Z_f, \; Z_{M2} = \frac{K}{1+K} Z_f. \tag{3.17}$$

By plugging the expressions for the capacitor impedances into the equations above, the formulae of Miller's theorem can be derived.

Figure 3.31(a) presents a CS amplifier circuit, akin to the CE amplifier illustrated in figure 3.24(a). During AC analysis within the high-frequency domain, external capacitors behave as short circuits. Upon replacing the MOSFET with its hybrid-π model, we obtain the small-signal circuit of this amplifier depicted in figure 3.31(b).

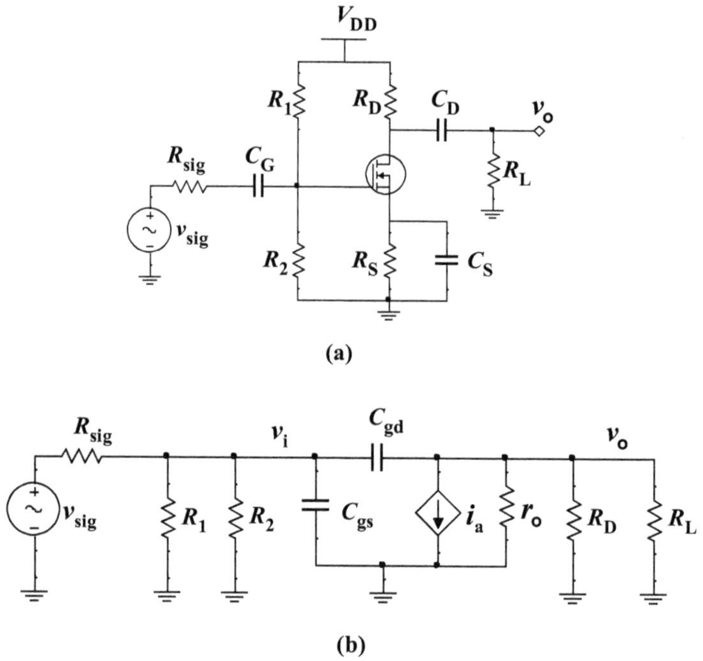

Figure 3.31. CS amplifier: (a) original amplifier circuit, (b) small-signal circuit.

Applying Miller's theorem, the capacitor C_{gd} can be converted into two capacitors, C_{m1} and C_{m2}, resulting in the transformed circuit shown in figure 3.32(a). The two capacitors on the input side can be consolidated, and the current source can be converted into a voltage source, leading to the circuit depicted in figure 3.32(b). Notably, given the interchangeable nature of parallel circuit elements, the subcircuit on the right-hand side actually shares the same configuration as the subcircuit on the left-hand side.

On the left-hand side of the circuit depicted in figure 3.32(b), Thévenin's theorem can be utilized, leading to the transformation of this subcircuit into an RC low-pass filter (LPF) circuit. Similarly, this approach can be employed for the subcircuit on the right-hand side. The high-frequency response is the result of cascading these two LPF circuits together. If the two poles are relatively close, the cutoff frequency can be determined using the following formula:

$$\frac{1}{f_H} = \sqrt{\frac{1}{f_{p1}^2} + \frac{1}{f_{p2}^2}}. \tag{3.18}$$

A common mistake in applying Miller's theorem is using the midband gain for the high-frequency domain. To illustrate this, an amplifier featuring a VCCS is designed, as shown in figure 3.33(a). With the amplifier's midband gain set at -100 V/V, C1 undergoes conversion into two capacitors: $C_{M1} = 101$ nF, $C_{M2} \approx 1$ nF. Calculating the frequencies of the resulting two poles yields $f_{p1} \approx 15.9$ kHz, $f_{p2} \approx 30.6$ kHz. When combined, these poles establish a cutoff frequency at $f_c \approx 14.1$ kHz.

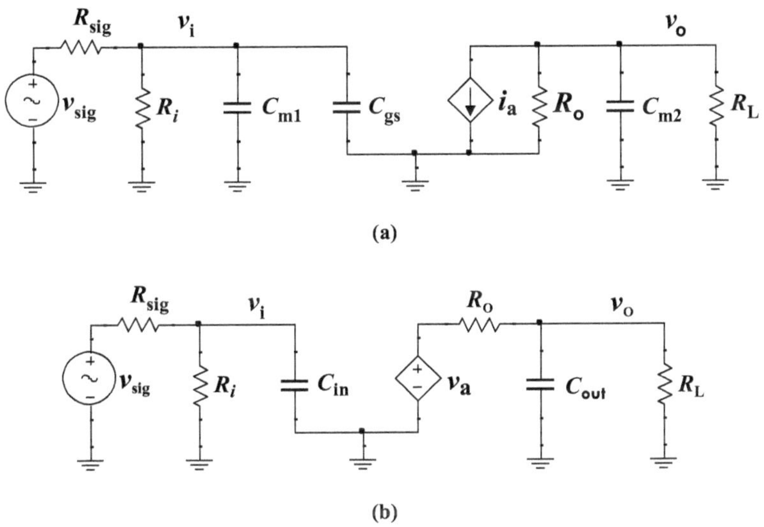

Figure 3.32. Equivalent circuits for a CS amplifier: (a) circuit obtained by applying Miller's theorem, (b) source transformation.

However, these findings contradict the Bode plot illustrated in figure 3.33(b). According to the simulation results, the cutoff frequency is about 10 kHz, exhibiting seemingly only one-pole behavior, as the Bode plot rolls off with a slope of −20 dB/dec.

Since the circuit shown in figure 3.33(a) is relatively simple, the expression for the gain can be directly derived using nodal analysis:

$$\begin{aligned} A_V(s) &= -g_m R \frac{1 - (C_1/g_m)s}{1 + \{[(1 + g_m R_L)R_s + R_L]C_1 + R_s C_2\}s + (R_s R_L C_1 C_2)s^2} \\ &= -g_m R \frac{1 - s/\omega_z}{\left(1 + s/\omega_{p1}\right)\left(1 + s/\omega_{p1}\right)}. \end{aligned} \quad (3.19)$$

While the first expression of equation (3.19) may appear intricate, substituting the circuit parameters changes its denominator into a quadratic equation. Solving this equation reveals the pole frequencies: $\omega_{p1} = 6.69 \times 10^4$ rad/s ($f_{p1} = 1.06 \times 10^4$ Hz) and $\omega_{p2} = 9.96 \times 10^6$ rad/s ($f_{p2} = 1.585 \times 10^6$ Hz). Furthermore, the zero frequency can be calculated directly: $\omega_z = 10^7$ rad/s ($f_z = 1.592 \times 10^6$ Hz). Given the proximity of the second pole and the zero frequencies, they effectively cancel each other out. Consequently, the frequency response is just like that of a first-order LPF with a cutoff frequency at 10.6 kHz. This outcome aligns closely with the simulation results.

Figure 3.34(a) shows the phase response of the Bode plot, which differs significantly from that of a typical single-pole system. The dominant pole at approximately 1.06×10^4 Hz introduces a standard −90° phase shift across the frequency range of 10^3–10^5 Hz. Normally, if the zero and the second pole effectively cancel each other, the phase curve should level off beyond 10^5 Hz, remaining flat.

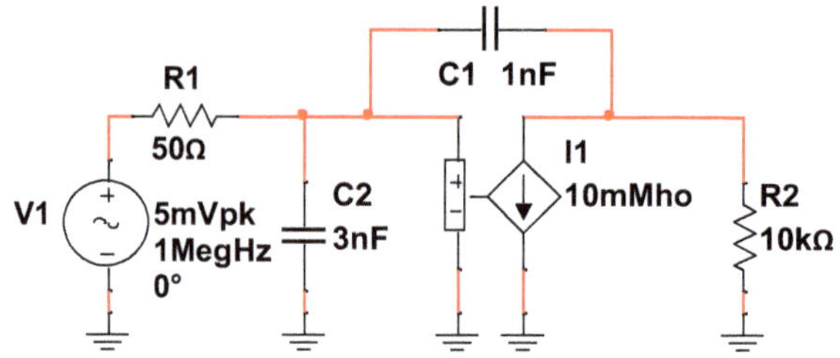

(a)

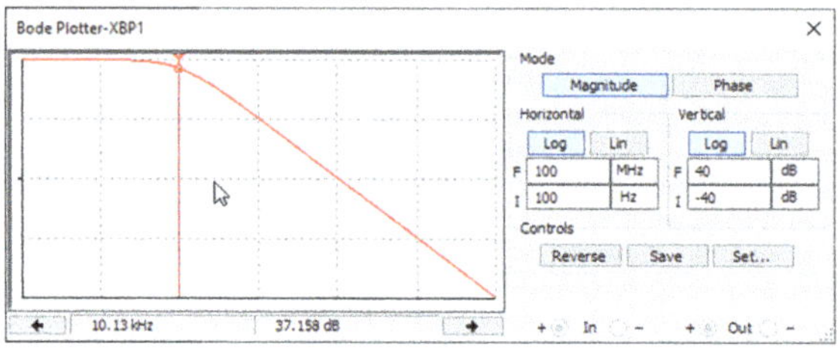

(b)

Figure 3.33. Investigation of Miller's theorem: (a) amplifier circuit, (b) Bode plot.

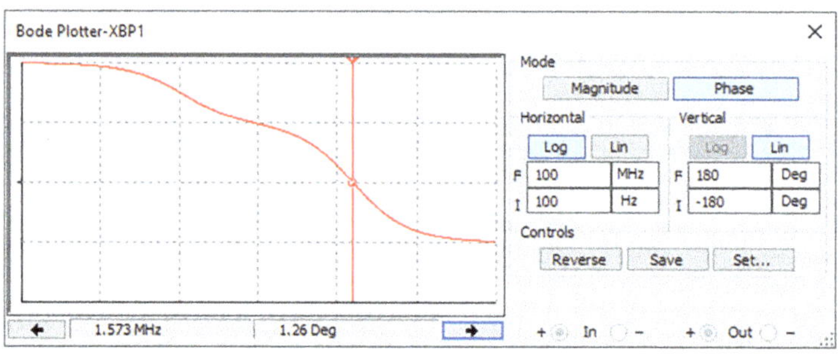

(a)

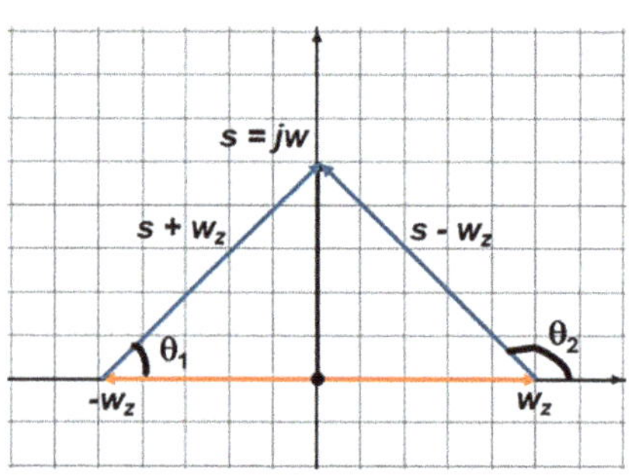

(b)

Figure 3.34. (a) Phase shift diagram, (b) pole diagram in the *s*-plane.

However, in this case, an additional $-180°$ phase shift is observed at around 1.59×10^6 Hz, suggesting an unexpected behavior that requires further analysis.

Referring to equation (3.19), we notice a negative sign in the numerator, indicating that the system includes a right half-plane (RHP) zero. This non-minimum phase element is responsible for the additional $-180°$ phase lag observed in the high-frequency range. To visualize this, figure 3.34(b) illustrates two real zeros located symmetrically on the s-plane at ω_Z and $-\omega_Z$. In addition, $s + \omega_Z$ and $s - \omega_Z$ are represented by two vectors with the phase angles of θ_1 and θ_2, respectively. The relationship between these two angles can be found from the symmetry: $\theta_2 = 180° - \theta_1$. If the direction of the vector $s - \omega_z$ is flipped, the phase angle of $\omega_Z - s$ becomes $\theta_3 = -\theta_1$, which doubles the phase shift of the second pole. Therefore, the $-90°$ phase shift becomes a $-180°$ phase shift.

> Miller's theorem is useful for understanding the amplification of the feedback capacitor. However, it is not very effective in determining the locations of the poles, and it also misses the zero in the transfer function. Therefore, one should be cautious when using it to analyze the high-frequency response of amplifier circuits.

3.9 CE/CS amplifiers with negative feedback

While high voltage gain is often the primary objective in amplifier design, it can sometimes be traded off to improve other parameters. In fact, all practical amplifiers utilize negative feedback mechanisms that can enhance stability, linearity, and bandwidth. Feedback in amplifiers is a major topic that is discussed further in chapter 5.

Figure 3.35(a) shows a CE amplifier with negative feedback, implemented by splitting the emitter degeneration resistor r_e into two parts. For DC analysis, these two resistors can be combined, making the circuit functionally identical to a traditional CE amplifier. For AC analysis, the top resistor (R_{E1}) provides negative feedback, as illustrated in figure 3.35(b).

To simplify the analysis, the source and load are removed. In addition, r_o is also ignored in the small-signal model. In this way, the voltage gain can be found easily:

$$i_a = \frac{\alpha v_i}{r_e + R_{e1}},\ v_o = -i_a R_C,\ A_{VO} = \frac{v_o}{v_i} = -\frac{\alpha R_C}{r_e + R_{E1}} \approx -\frac{g_m R_C}{1 + g_m R_{E1}} \tag{3.20}$$

Next, we can find the input resistance using the *resistance reflection rule*:

$$R_i = (\beta + 1)(r_e + R_{E1}) = r_\pi + (\beta + 1)R_{E1} \approx r_\pi(1 + g_m R_{E1}). \tag{3.21}$$

Compared with the circuit without a partitioned R_E, the voltage gain of this circuit is lowered by a factor of $1 + g_m R_{E1}$, and the input resistance is increased by the same factor of $1 + g_m R_{E1}$. This common factor is known as the ***amount of feedback***. In summary, splitting R_E into two parts increases the input resistance and reduces the voltage gain. Additionally, feedback also improves linearity and bandwidth.

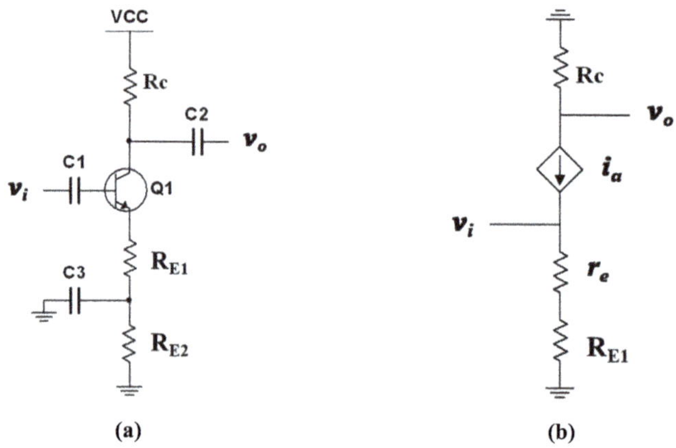

Figure 3.35. Amplifier with negative feedback: (a) core amplifier circuit, (b) small-signal circuit.

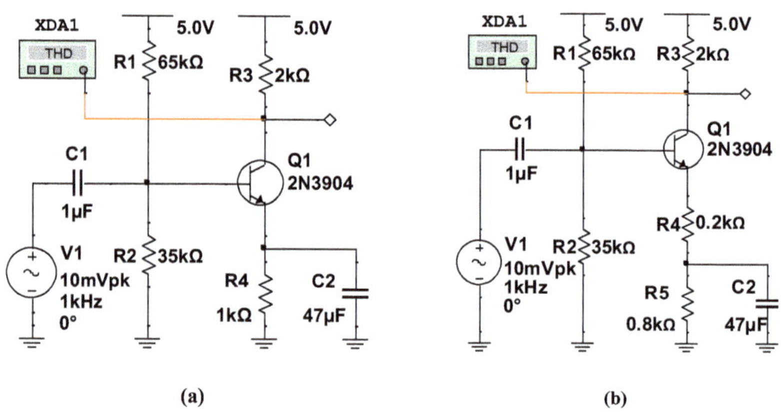

Figure 3.36. CE amplifier circuit: (a) without feedback, (b) with feedback.

To illustrate the improvement in linearity, the resistor R4 in figure 3.36(a) is split into two resistors, resulting in the circuit depicted in figure 3.36(b). Distortion can be analyzed using Fourier series, which transform the output signal into a sum of fundamental and harmonic components. Instead of examining all these harmonics individually, they can be combined into a single parameter known as total harmonic distortion (THD), which is defined as follows:

$$THD = \frac{\sqrt{V_2^2 + V_3^2 + \cdots}}{V_1}, \tag{3.22}$$

where V1 is the amplitude of the fundamental frequency and V2, V3, ... are the amplitudes of the harmonics. Multisim offers a virtual instrument for measuring THD, making the simulation process very convenient.

The simulation results are shown in figure 3.37, indicating a substantial improvement in linearity. In other words, the distortion is reduced significantly. It is

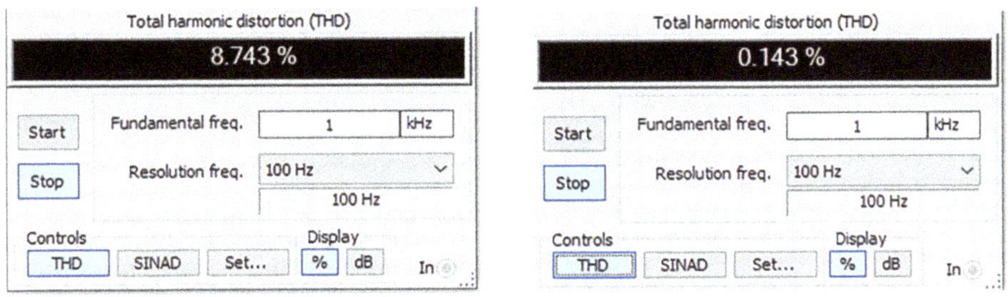

(a) (b)

Figure 3.37. Comparison of THD: (a) without feedback, (b) with feedback.

important to note that the THD value depends on the amplitude of the input signal. For example, if the amplitude of the input signal is reduced to 5 mV, the THDs for the circuits decrease to 4.382% and 0.071%, respectively.

> The acceptable level of THD varies across different applications. For example, a THD below 1% is required for consumer audio equipment, a THD below 0.5% is needed for professional audio equipment, and a THD below 0.1% is necessary for high-fidelity audio equipment. In the power electronics field, the requirements are less stringent, and the THD is required to be below 5%.

3.10 CB/CG amplifier circuits

The CB configuration is widely used in multistage amplifiers and ICs. However, it is not ideal for single-stage voltage amplifiers due to its very low input resistance. Figure 3.38(a) presents an example of a core CB amplifier whose DC section is identical to that of the CE amplifier shown in figure 3.20(a). As before, we can assume all capacitors are large enough that they disappear in the small-signal circuit, which is illustrated in figure 3.38(b). Additionally, R_{B1} and R_{B2} are absent, since both ends are grounded in AC.

This circuit can be analyzed using the same procedure as that used for the CE amplifier. The voltage gain of the core amplifier can be derived similarly:

$$i_a = g_m v_{be} = -g_m v_i,\ \ v_o = -i_a R_C = g_m R_C v_i,\ \ A_{VO} = \frac{v_o}{v_i} = g_m R_C. \tag{3.23}$$

Interestingly, the magnitude of the gain is the same as that of the core CE amplifier, but its sign is opposite. This can be explained by examining the small-signal circuit in figure 3.38(b). In the CE amplifier circuit, the top of r_e is connected to the input signal, and the bottom is grounded. In the CB amplifier circuit, this configuration is reversed, and thus the voltage gain becomes positive.

The output resistance of this amplifier is the same as that of the CE amplifier, provided r_o can be ignored. However, the input resistance is quite different. From

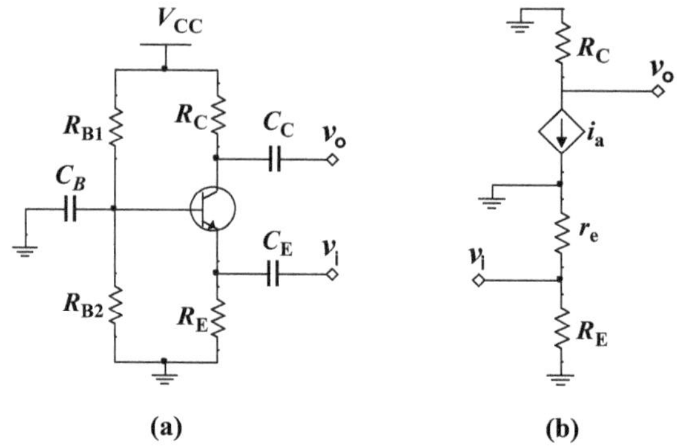

Figure 3.38. CB BJT amplifier: (a) original circuit, (b) small-signal circuit.

the perspective of the input signal, r_e and R_E are in parallel, but typically the former is much smaller:

$$R_o = R_C, \ R_i = R_E // r_e \approx r_e. \tag{3.24}$$

Taking the source resistance and load resistance into account, the gain of this amplifier circuit can be derived in the same way as for the CE amplifier:

$$A_V = \frac{v_o}{v_{sig}} = \frac{R_i}{R_{sig} + R_i} g_m R_o \frac{R_L}{R_o + R_L} \tag{3.25}$$

Although this expression is very similar to that of the CE amplifier, the voltage gain is much lower due to the very low input resistance (R_i). Additionally, the CB amplifier does not provide any current gain. Furthermore, if the internal resistance of a voltage signal source is quite high, the CB configuration is not a good choice. On the other hand, if the signal source is a current source in parallel with an internal resistance, then the low input resistance of the CB amplifier becomes an advantage. This can be understood using the analogy of a current divider circuit, where most of the current from the source goes through the branch with the lower resistance.

Compared to CE amplifiers, CB amplifiers offer much higher bandwidth, making them well-suited for high-frequency applications. One key reason is that the voltage gain of a CB amplifier is positive, so Miller's theorem—which significantly increases the equivalent input capacitance in inverting amplifiers—does not apply. Furthermore, the internal collector–emitter capacitance C_{CE} is typically very small, resulting in minimal feedback and reduced phase distortion at high frequencies. Figure 3.39 presents the simulation results confirming these characteristics. First, the input and output signals are in phase, as shown at the top of figure 3.39(b), which is consistent with the non-inverting nature of the CB configuration. Second, the midband voltage gain reaches 25.4 V/V (28.1 dB), as illustrated in figure 3.39(a). Third, and most notably, the amplifier achieves a bandwidth of 26.2 MHz, as shown at the bottom of figure 3.39(b)—an order of magnitude greater than that of comparable CE amplifiers.

Figure 3.40(a) shows a CG amplifier with a current source placed above a negative DC voltage source, allowing the gate to be directly grounded. The voltage

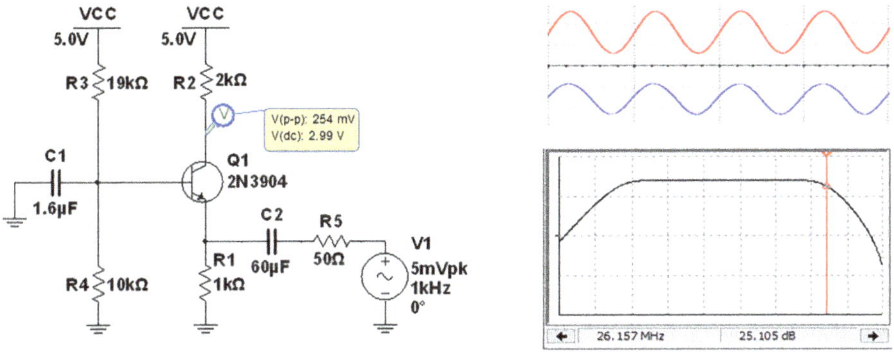

(a) (b)

Figure 3.39. CB BJT amplifier: (a) circuit, (b) simulation results.

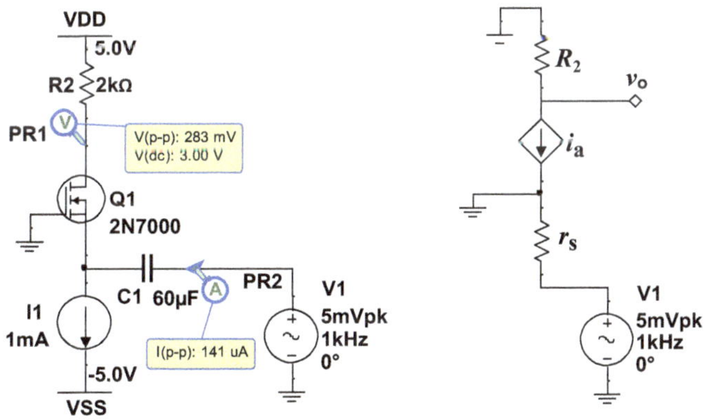

(a) (b)

Figure 3.40. CG MOSFET amplifier: (a) original circuit, (b) small-signal results.

gain can be obtained directly from the simulation results: $A_V = 28.3$ V/V. From this value, the transconductance is derived: $g_m = 14.2$ mS. This value is lower than that of a BJT ($g_m = 25.9$ mS) with the same current. The input resistance r_s can be found from the simulated input current, as shown in the small-signal circuit in figure 3.40(b): $r_s \approx 70.5\ \Omega$. This resistance can also be indirectly calculated from the transconductance, $r_s = 1/g_m$. Compared to the CB BJT amplifier with the same transistor current, r_s is larger than r_e, which is 25.9 Ω at a current of 1 mA.

> Impedance matching is crucial in RF and microwave circuits to ensure maximum power transfer and minimize signal reflection. Coaxial cables typically have impedances of either 50 Ω (lab equipment) or 75 Ω (cable TV), requiring the input impedances of the amplifiers to be close to these values. Therefore, the CB amplifier, with its low input impedance, is suitable for use in these applications.

3.11 CC/CD amplifier circuits

The CC amplifier is a popular configuration in electronic circuits. It consists of a transistor where the input signal is applied to the base, the output is taken from the emitter, and the collector is typically connected to a DC voltage source. An example is shown in figure 3.41(a). This configuration is widely used due to its high input impedance, low output impedance, and approximately unity voltage gain, making it an excellent buffer stage or output stage. Unlike CE and CB amplifiers, CC amplifiers can also work with large input signals without causing severe distortion.

This type of amplifier is also known as an *emitter follower*, indicating that the output signal at the emitter follows the input signal at the base. This behavior is due to the assumption that the base–emitter voltage is constant, so the AC signals at the base and emitter are approximately the same. However, the voltage gain is slightly less than unity, as demonstrated by the simulation results ($A_V \approx 0.972$ V/V) in figure 3.41(a). This slight reduction in gain is expected, since small variations in v_{BE} occur when an input signal is applied. As illustrated in the small-signal circuit shown in figure 3.41(b), a voltage divider is present at the emitter, causing a small portion of the input signal to drop across the base–emitter junction, represented as r_e in this small-signal circuit. In addition, the source resistance R_1 is ignored, which will be discussed in more detail.

When the input signal is not weak, the small-signal model does not work. In this case, we need to rely on the *I*–*V* characteristics of the *pn* junction: a large change in the current corresponds to a small change in the voltage. Therefore, most of the input signal still falls on the load resistor, so the gain should remain close to unity. However, due to the nonlinearity of the *I*–*V* curve, the output waveform can be distorted when the input signal is too strong.

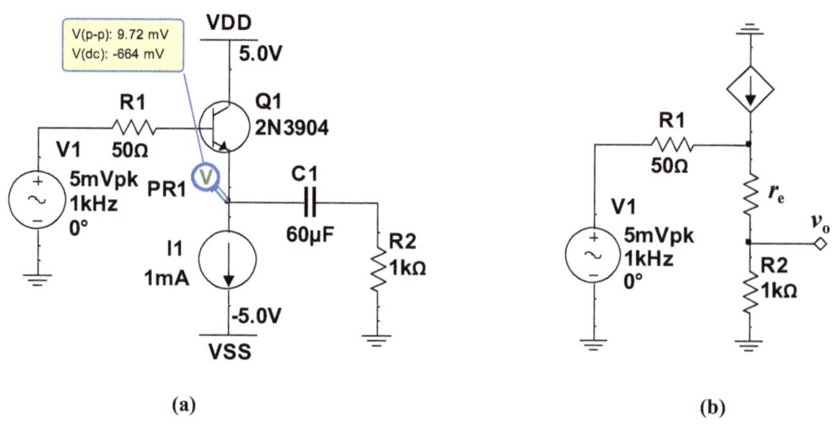

Figure 3.41. CC amplifier: (a) original circuit, (b) small-signal circuit.

Although the small-signal circuit shown in figure 3.41(b) appears quite simple at first glance, determining its input and output resistances can be surprisingly challenging. A useful technique for analyzing this type of circuit is the *resistance reflection rule*, which is especially effective in circuits involving current gain, such as those using BJTs. In this configuration, the current flowing through R_1 is i_b, while the current passing through r_e and R_2 is i_e, and they are related by $i_e = (\beta + 1)i_b$. This current gain impacts the way resistances are reflected between terminals. Specifically, any resistance in the emitter leg—such as the resistances r_e and R_2 —is reflected into the base circuit as a larger resistance by a factor of $(1 + \beta)$. Therefore, the input resistance seen at the base is significantly increased due to this reflection effect. Similarly, the output resistance, which depends on how the base resistance reflects into the emitter or collector side, should also consider this current relationship. Including these transformations in the analysis allows for accurate expressions of both the input and output resistances:

$$\begin{aligned} R_i &= (\beta + 1)(r_e + R_2) = r_\pi + (\beta + 1)R_2 \\ R_o &= r_e + R_1/(\beta + 1). \end{aligned} \tag{3.26}$$

Assuming $\beta = 149$, the input resistance is $R_i \approx 154\ \text{k}\Omega$ and the output resistance is $R_o \approx 26.2\ \Omega$. These values highlight the characteristics of CC amplifiers, namely high input impedance and low output impedance, which make them effective as a buffer between two amplifier stages.

The voltage gain can be derived from the voltage divider formula. From the perspective of the signal source, the two resistors (r_e and R_2) at the emitter node are effectively boosted by the current ratio:

$$A_V = \frac{v_o}{v_{sig}} = \frac{(\beta + 1)R_2}{R_1 + (\beta + 1)(r_e + R_2)}. \tag{3.27}$$

Usually, $R_2 > > r_e$, and $(\beta + 1)R_L > > R_1$, so the voltage gain is close to unity.

> If the input resistance needs to be further increased, the BJT can be replaced with a Darlington pair, which is a combination of two BJTs. The current gain of a Darlington pair is approximately the product of the gains of these two individual BJTs, resulting in a very high overall current gain.

Figure 3.42(a) shows a CC amplifier without a current source, making it easy to implement in the lab. A large input signal with an amplitude of 250 mV is applied, and the distortion is measured at the output node, which is found to be at an acceptable level of 4.15% for power electronics.

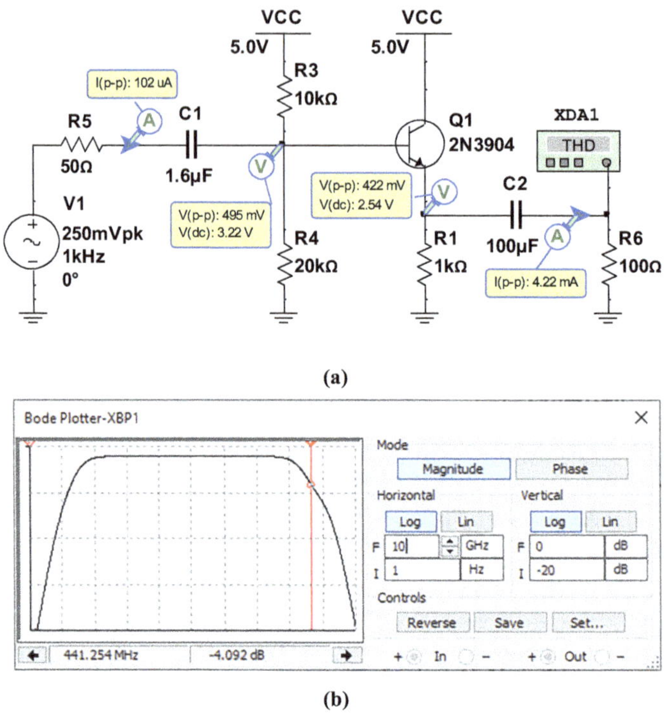

Figure 3.42. CC amplifier: (a) circuit, (b) Bode plot.

The simulation results in figure 3.42 reveal the following parameters:

- Voltage gain: $A_V = 0.844$ V/V
- Current gain: $A_I = 41.4$ A/A
- Bandwidth: BW = 441 MHz

Although the voltage gain of CC amplifiers is less than unity, they have relatively high current gain, making them suitable for use as output stages or power amplifiers. Additionally, CC amplifiers offer very high bandwidth, typically an order of magnitude higher than that of CB amplifiers. This increased bandwidth is due to the deep negative feedback inherent in CC amplifiers, which extends the bandwidth significantly.

Figure 3.43 shows the simulation results of a CD amplifier with a MOSFET, which can be analyzed similarly to the CC amplifier shown in figure 3.41(a). With an input signal amplitude of 500 mV, the distortion of the output signal is still quite low: THD = 2.7%. A key advantage of MOSFETs is their very high input impedance and very low gate current, as demonstrated in the simulation results ($I_{p-p} = 168$ nA). From the voltage gain of 0.746 V/V, the output impedance can be derived, namely $R_o \approx 34\ \Omega$. Therefore, CD amplifiers share the same characteristics as CC amplifiers: high input impedance, low output impedance, high bandwidth, and low distortion with large input signals.

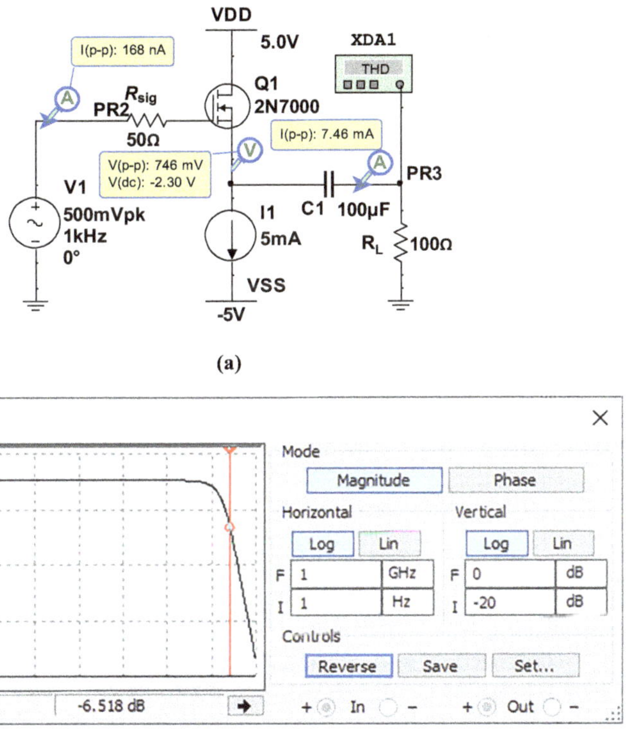

(a)

(b)

Figure 3.43. CD amplifier: (a) circuit, (b) Bode plot.

CC/CD amplifiers, also known as emitter/source followers, are widely used in electronics due to their unique characteristics. They provide high input impedance, low output impedance, and a voltage gain of approximately unity. These properties make them ideal for several applications, including impedance matching, buffering stages, and voltage regulation. By effectively isolating different stages of a circuit, they ensure minimal signal loss and distortion, making them indispensable in audio amplifiers, signal processing, and power supply circuits. Additionally, their capability to drive low-resistance loads without a substantial voltage drop extends their usefulness in both analog and mixed-signal ICs, including sensor interfaces, Op-Amp output stages, and digital-to-analog converters (DACs).

Chapter 4

Differential amplifier circuits

A differential amplifier (DA) amplifies the voltage difference between two input signals while rejecting any voltage common to both. This unique property makes it an essential building block in analog circuit design, particularly in applications that demand high precision and noise immunity. Key performance attributes of DAs include a high common-mode rejection ratio (CMRR), high input impedance, and the ability to process differential signals. These characteristics make them indispensable in operational amplifiers, analog-to-digital converters, and various signal-conditioning circuits, where accuracy, stability, and robustness against interference are critical.

In the fabrication of integrated circuits (ICs), unavoidable process variations lead to slight mismatches in transistor characteristics, often deviating from ideal design parameters. Such variations can cause shifts in the DC operating point in amplifier circuits like those discussed in the previous chapter. While a single-stage amplifier can typically tolerate these shifts, multistage configurations—such as those used in operational amplifiers—are far more sensitive. Voltage offsets can accumulate across stages, degrading performance. DAs offer a solution to this problem: they rely on the matching of transistor pairs rather than absolute parameter values. Since matched devices can be closely placed and fabricated under identical conditions on a chip, differential architectures are inherently more robust and better suited for integration in high-performance analog systems.

This chapter introduces the fundamental principles and practical implementations of DAs, beginning with their basic configuration and operation. It then explores the critical role of current mirrors—both simple and advanced types—in biasing and active loading. The discussion extends to discrete amplifier circuits with active loads, followed by a detailed examination of differential pairs enhanced with active load techniques. Cascode amplifiers are introduced as an effective strategy to improve gain and bandwidth, and the chapter concludes with an overview of multistage DAs, which are widely used in high-performance analog systems such as

doi:10.1088/978-0-7503-5512-4ch4

operational amplifiers and data converters. Through theoretical analysis and practical examples, this chapter builds a comprehensive understanding of DA architectures and their essential building blocks.

4.1 Introduction to differential amplifiers

The core of a DA is a pair of matched transistors, as illustrated in figure 4.1. Unlike the amplifier circuits discussed in the previous chapter, a DA has two input signals connected to the base/gate nodes of the transistors. The output signal is taken from the collector/drain nodes, similar to a CE/CS amplifier configuration. There are two different output modes:

- Single-ended output modes: the signal is taken from one side only, $v_o = v_{o1}$ or $v_o = v_{o2}$.
- Fully differential output modes: the signal is taken from both sides, $v_{od} = v_{o2} - v_{o1}$.

The operating principle of a DA is based on competition for current. As shown in figure 4.1, the sum of the currents from the two branches is constant. The current distribution is controlled by the so-called 'differential input signal': $v_d = v_{i1} - v_{i2}$. For a positive differential input signal ($v_d > 0$), more current flows through the left branch, and vice versa for a negative differential input signal ($v_d < 0$).

Figure 4.2 illustrates the current distribution in a BJT DA. The solid line represents the current in the right branch, while the dashed line represents the current in the left branch. First, the sum of these two curves remains constant for any differential input signal, in accordance with Kirchhoff's current law (KCL). Second, the current distribution is highly sensitive to the differential input signal; a 0.1 V difference between the two input signals can shift almost all the current to one side. A similar phenomenon occurs in MOSFET DAs, though their sensitivity to the differential input signal is slightly different.

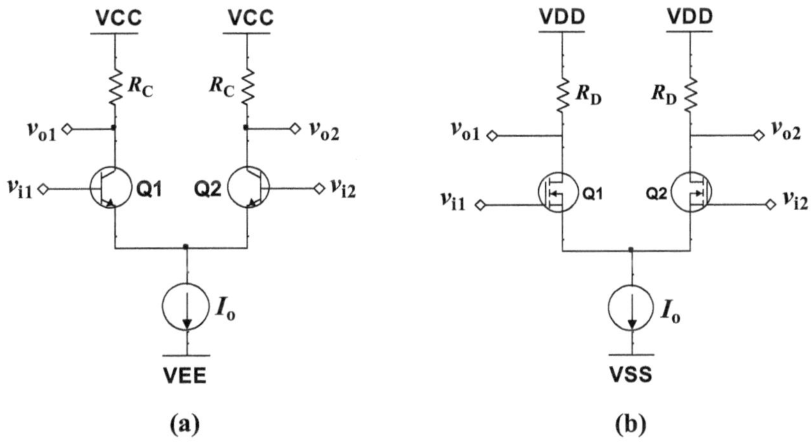

Figure 4.1. DA circuits with an ideal current source: (a) BJT circuit, (b) MOSFET circuit.

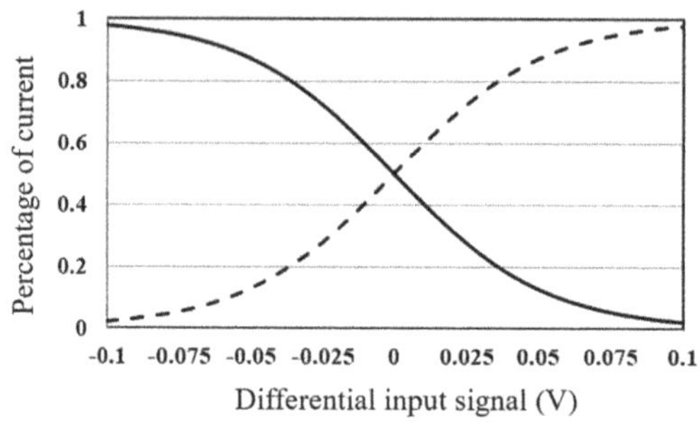

Figure 4.2. Current switching in a BJT DA.

When the differential-mode input signal is strong, this circuit operates in 'digital mode,' with all the current directed to either the left or the right side. When implemented using BJTs, this type of logic circuit is known as 'emitter-coupled logic (ECL),' which was widely used in supercomputers from the 1960s to the 1980s because of its high performance. Unlike the BJT inverter circuit discussed in the previous chapter, ECL does not suffer from the delay associated with BJTs in saturation mode. In other words, the BJTs in these circuits are either in cutoff mode or active mode; they are never in saturation mode.

> ECL was invented in 1956 and gained popularity in the 1960s when BJTs dominated electronics. However, ECL has a significant drawback: very high power consumption. Due to the current source at the bottom, the power consumption remains high even when no logic operation is performed. In contrast, CMOS logic circuits do not consume power in this idle state. Additionally, CMOS circuits are cheaper to fabricate and have a higher density. As a result, ECL became obsolete in the late 1980s with the widespread adoption of MOSFETs for ICs.

When the differential input signal is weak, the circuit works as an amplifier using the principle described in the previous chapter: first, the input voltage signal is converted into a current signal by the transistor pair. Second, this current signal is converted back into an output voltage signal by the resistors at the top. This can be described in a figurative way: the input voltage signal causes a swing of the current between the two branches; then, the voltages at the output nodes swing according to Ohm's law.

As illustrated in figure 4.3, any two input signals, represented by the two columns, can be expressed as the superposition of a 'common-mode' input signal (v_{cm}) and a 'differential-mode' input signal (v_{d}):

$$v_{\mathrm{i1}} = v_{\mathrm{cm}} + \frac{1}{2}v_{\mathrm{d}},\ v_{\mathrm{i2}} = v_{\mathrm{cm}} - \frac{1}{2}v_{\mathrm{d}}. \tag{4.1}$$

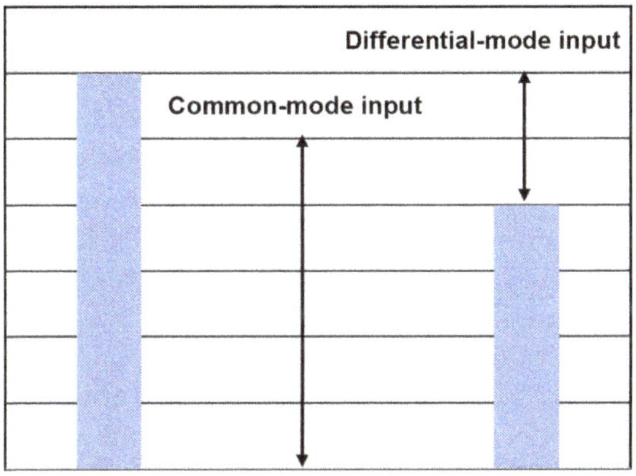

Figure 4.3. Common-mode and differential-mode signals.

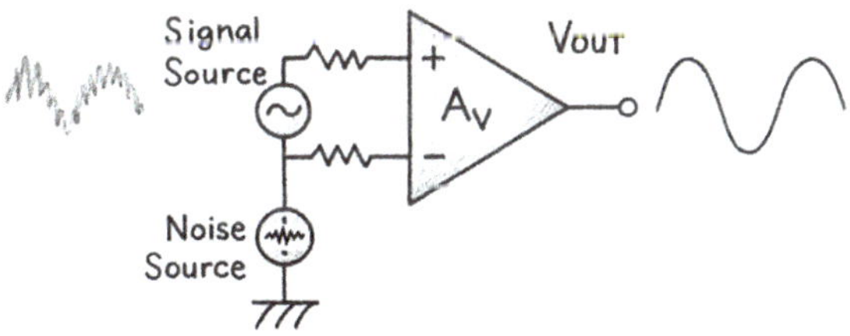

Figure 4.4. Common-mode and differential-mode signals in an amplifier circuit. Created with GPT-4.0, OpenAI.

On the other hand, these two new parameters (v_{cm} and v_d) can be derived from the original input signals: the common-mode signal is the average of the two input signals, while the differential-mode signal is simply the difference between them:

$$v_{cm} = \frac{1}{2}(v_{i1} + v_{i2}), \ \ v_d = v_{i1} - v_{i2}. \tag{4.2}$$

In many practical applications, the differential-mode input signal carries weak but valuable information that needs amplification. Conversely, the common-mode signal often contains strong interference and noise that must be rejected. As illustrated in figure 4.4, the signal source at the top left represents the useful signal from a sensor, while the noise signal source below it represents the noise and interference. Therefore, the two wires leading to the amplifier carry identical interference and noise signals, which can be much stronger than the useful differential-mode signal. For example, if these wires are relatively long, they can effectively function as antennas, picking up various interference signals.

In general, both the differential-mode and common-mode signals appear at the output of amplifiers, but the gains for these two modes (A_d and A_{cm}) are quite different:

$$v_o(t) = A_d v_d(t) + A_{cm} v_{cm}(t). \tag{4.3}$$

In many situations, the output signal is taken from one side. To specify the characteristics of DAs, a figure of merit known as the CMRR is defined as follows:

$$CMRR = 20\log_{10}\left|\frac{A_{\mathrm{d}}}{A_{\mathrm{cm}}}\right|. \tag{4.4}$$

For good DAs, the differential-mode gain (A_{d}) is much higher than the common-mode gain (A_{cm}); therefore, CMRR is conveniently represented in dB. For example, $A_{\mathrm{d}} = 100$ V/V, $A_{\mathrm{cm}} = -0.01$ V/V, and then CMRR = 80 dB.

The DA circuits shown in figure 4.1 can reject the common-mode input signal completely ($A_{\mathrm{cm}} = 0$) because an ideal current source is at the bottom and the circuit is perfectly symmetric. This can be understood using an analogy: In a small town with two thousand voters, two political parties are neck-and-neck in competition. If all the voters cast their votes, this situation is equivalent to the ideal current source in the circuit, which provides a constant current. In this case, each party receives one thousand votes (equal currents in the two branches). If both parties adopt the same campaign measures (common-mode input), the number of votes received by each party remains unchanged, rendering these efforts useless (common-mode rejection). However, if some voters are initially reluctant to vote but are motivated by these campaigns, then each party can receive more votes, analogous to the common-mode gain. In the following sections, the ideal current source is replaced by practical devices, such as resistors and transistors, resulting in a nonzero common-mode gain.

With the common-mode input signal rejected completely, the differential-mode input can be implemented using a single signal source, as illustrated in figure 4.5(a). In this circuit, $v_{\mathrm{i1}}(t) = v_{\mathrm{d}}(t)$ and $v_{\mathrm{i2}}(t) = 0$, and the common-mode input signal can be derived from equation (4.2): $v_{\mathrm{cm}}(t) = 0.5v_{\mathrm{d}}(t)$. The waveform displayed in figure 4.5(b) indicates a 180° phase shift between the output signals on both sides, reflecting that a portion of the current is swinging back and forth from one side to

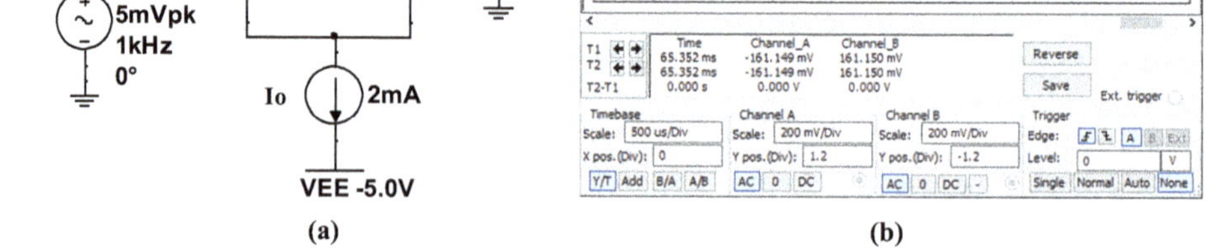

Figure 4.5. (a) DA circuit with an ideal current source, (b) simulation result.

the other. The differential-mode voltage gain can be obtained from the simulation results: $A_\mathrm{d} = \pm 40.2\ \mathrm{V/V}$.

4.2 Basic differential amplifiers

The DA circuit shown in figure 4.1 requires an ideal current source, which is unfortunately impractical in real-world applications. A simple alternative is to replace it with a resistor, but this results in suboptimal performance. A more effective solution is to use a transistor, which can act both as a current source and a resistor. In the next section, we explore more advanced current source designs.

Figure 4.6(a) shows a primitive DA circuit with a tail resistor R_E. In analyzing this circuit, the common mode and differential mode can be treated separately. First, assume the input signal is in pure differential mode with a small magnitude: $v_\mathrm{i1} = 0.5v_\mathrm{d}$ and $v_\mathrm{i2} = -0.5v_\mathrm{d}$. In this case, the emitter node is pulled up from one side and pulled down from the other. Therefore, it is reasonable to assume that the voltage at this node remains unchanged. In other words, it is equivalent to AC ground, transforming this circuit into two common-emitter (CE) amplifiers. Using the formula for CE amplifier gain, the output signals can be determined as follows:

$$\begin{aligned} v_\mathrm{o1} &= -\frac{1}{2}g_\mathrm{m}(R_\mathrm{C} \| r_\mathrm{o})v_\mathrm{d} \approx -\frac{1}{2}g_\mathrm{m}R_\mathrm{C}v_\mathrm{d} \\ v_\mathrm{o2} &= \frac{1}{2}g_\mathrm{m}(R_\mathrm{C} \| r_\mathrm{o})v_\mathrm{d} \approx \frac{1}{2}g_\mathrm{m}R_\mathrm{C}v_\mathrm{d}. \end{aligned} \tag{4.5}$$

An approximation is made in these formulae by assuming $r_\mathrm{o} \gg R_\mathrm{C}$, which is valid for discrete transistors. In addition, these formulae can be applied to DA circuits with any devices or subcircuits below the transistor pair. Therefore, they can also be used in the circuit shown in figure 4.5(a), where the simulation results indicate that the voltage gain is 40.2 V/V. With a 2 mA current source at the bottom, the

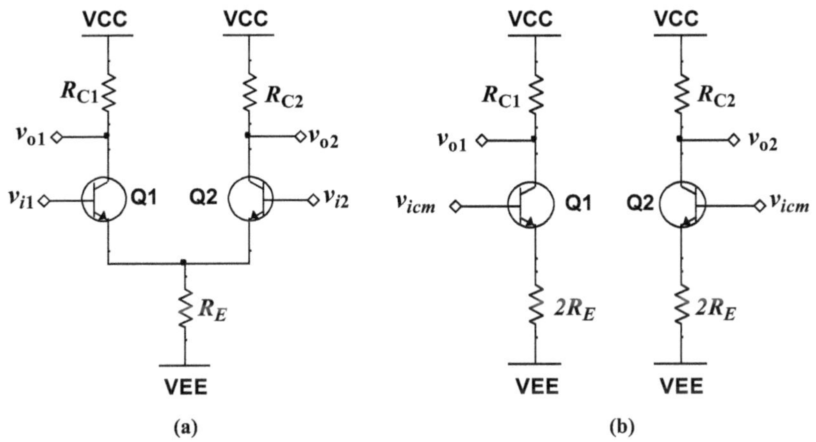

Figure 4.6. (a) Cheap DA, (b) its half-circuits with a common-mode input.

transconductance for each transistor is $g_m \approx 38.6$ mS, so the voltage gain calculated using equation (4.5) is 42.5 V/V. The discrepancy arises mainly from the approximation made in ignoring the output resistance (r_o).

If the differential-mode input is turned off, the symmetry of this circuit allows us to split it into two identical half-circuits, as shown in figure 4.6(b). In this transformation, the value of the tail resistor is doubled, since the combination of these two resistors in parallel needs to match the original resistor. With a resistor below the emitter, the T-model is preferred in AC analysis, and the voltage gain can be derived as follows:

$$v_{o1} = v_{o2} = -\frac{\alpha R_C}{2R_E + r_e} v_{cm} \approx -\frac{R_C}{2R_E} v_{cm}. \quad (4.6)$$

Keep in mind that this common-mode gain is undesirable, so the lower it is, the better. This equation indicates that the gain is inversely proportional to R_E, so it seems that a very large resistor can be used here. However, Ohm's law limits the range of this resistor. If it becomes too large, the DC current needs to be low. Consequently, the transconductance of the transistors (g_m) decreases, and the differential-mode gain suffers. Thus, both the common-mode and differential-mode gains need to be considered simultaneously, as reflected in the CMRR:

$$CMRR = 20 \log_{10}\left|\frac{A_d}{A_{cm}}\right| \approx 20 \log_{10}(g_m R_E). \quad (4.7)$$

For example, if $R_E = 2.2$ kΩ and $I_{RE} = 2$ mA, then $I_C \approx 1$ mA, $g_m \approx 38.6$ mS, and CMRR ≈38.6 dB. With a resistor at the tail, the CMRR cannot improve further. For instance, if the DC level of the input signal is zero ($V_B = 0$ V), then $V_E \approx -0.7$ V. In addition, $I_{RE} R_E = V_E - V_{EE}$, and $g_m R_E = I_C R_E / V_T \approx (V_E - V_{EE})/(2V_T)$, which is independent of R_E.

The formula for the CMRR is typically defined for single-ended output configurations. However, in differential output mode, common-mode signals can still be effectively rejected—particularly when the circuit is symmetric. Even in the presence of slight mismatches between the transistor pair or the collector resistors, the resulting common-mode output signal remains minimal. Equation (4.8) quantifies the gain from the common-mode input to the differential output:

$$v_{od,cm} = v_{o2,cm} - v_{o1,cm} \approx \frac{g_{m1} R_{C1} - g_{m2} R_{C2}}{2\bar{g}_m R_E} v_{cm}. \quad (4.8)$$

If the transistors and collector resistors are well matched, the differential output due to a common-mode input becomes negligibly small, resulting in a very high CMRR. This excellent rejection of common-mode interference is one of the key advantages of the differential output mode and is crucial for high-precision analog applications, such as low-noise sensor interfaces and differential signal processing. The effectiveness of this configuration underscores the importance of careful layout and matching in IC design, especially for sensitive analog front-end circuits.

In circuit diagrams, the pair of transistors is often shown in a *mirrored* arrangement. However, on a circuit board, it is impossible to place two BJTs in this way, as the emitter and collector pins would be reversed. Therefore, one cannot strictly follow the circuit diagram when constructing the circuit; instead, the two transistors should be placed in *parallel*. In contrast, low-power MOSFETs are symmetric devices, so they can be flipped. However, power MOSFETs are asymmetric, as they have a different device structure from that of low-power MOSFETs.

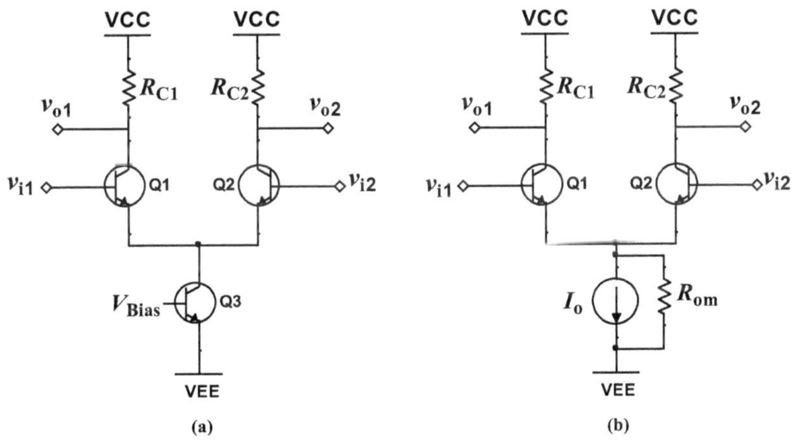

Figure 4.7. (a) Simple DA, (b) its equivalent circuit.

To increase the CMRR, both the tail current and the tail resistor R_E must be large, and this can be achieved by replacing R_E with a transistor, as shown in figure 4.7(a). Here, the transistor is effectively a combination of a current source I_o and an output resistance R_{om}, and this equivalent circuit is shown in figure 4.7(b). This transformation can be understood directly from the I–V characteristics of transistors in active mode. Due to the Early effect, this curve has a small slope, which can be modeled by a resistor R_{om} in parallel with the current source, just like the hybrid-π small-signal model. If a single transistor is used, as in the circuit shown in figure 4.7(a), the output resistance is $R_{om} = r_o$.

As discussed earlier, the analyses of the differential mode and the common mode can be separated. In the differential-mode analysis, the CE node of the transistor pair can still be assumed to be at AC ground, so the formula for the gain in equation (4.5) is still valid: $A_d = \pm 0.5 g_m R_C$. Since $g_m = I_C/V_T$ and $A_d = \pm\, 0.5 I_C R_C/V_T = \pm\, 0.5\ V_R/V_T$, the magnitude of this gain is limited by the DC voltage across the resistor R_C.

In DA design, careful attention must be paid to the DC collector voltage of the transistors. If this voltage drops too low, the transistors may enter saturation, causing the amplifier to fall outside its linear operating region and suffer from degraded performance. The collector voltage is primarily determined by the collector current, which in turn is set by the tail current flowing through the transistor connected to the CE node—often referred to as the *tail transistor*. This current is controlled by the bias voltage (V_{Bias}) applied to its base. Precise adjustment of V_{Bias} is necessary to ensure that the transistors operate in the active region and maintain proper headroom. However, in IC design, such fine-tuning is impractical due to process variations and limited access to individual bias points. To address this challenge, the next section introduces the current mirror, a widely used circuit block that provides a stable and predictable tail current, improving reliability and simplifying biasing in DA stages.

In the AC analysis of circuits with a common-mode input, the DC current source I_o can be removed from the circuit, so the circuit shown in figure 4.7(b) is the same as the circuit in figure 4.6(b). Therefore, the common-mode gain is also the same, as well as the CMRR. The transconductance of the transistor pair is $g_m = \alpha I_o/(2V_T)$, and the output resistance of the tail transistor is related to the Early voltage: $R_{om} = r_o = V_A/I_o$, so the CMRR can be found:

$$\text{CMRR} \approx 20\log_{10}(g_m R_{om}) \approx 20\log_{10}\left(\frac{V_A}{2V_T}\right). \tag{4.9}$$

With the tail resistor R_E replaced by a transistor, the CMRR is increased significantly. For example: $V_A = 100$ V, $V_T = 25.9$ mV, and CMRR $\approx$ 66 dB. However, equation (4.9) indicates that the CMRR is limited by the Early voltage of the tail transistor. To increase the CMRR further, the tail transistor can be replaced by an advanced subcircuit, which is discussed in the next section.

Figure 4.8 illustrates a DA circuit with a common-mode input signal and the simulated waveform at the output nodes. First, the gain can be determined from the simulation results: $A_{cm} = -0 \cdot 025$ V/V. From this value, the output resistance of the tail transistor can be estimated using equation (4.6): $r_{o3} \approx 44$ kΩ, and the Early

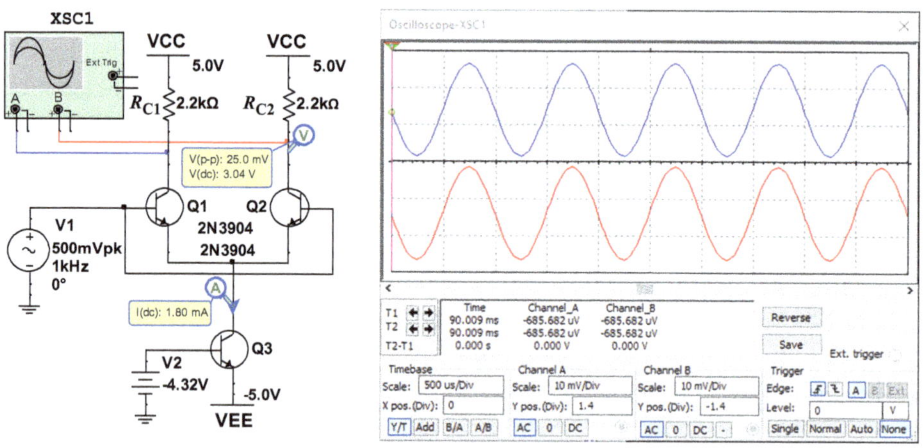

(a) (b)

Figure 4.8. (a) DA circuit with a common-mode input, (b) waveforms of output signals.

voltage can be derived: $V_{A3} \approx 79$ V. In addition, the CMRR can be calculated using equation (4.9): $CMRR \approx 63.7$ dB. Second, the bias voltage at the base of the tail transistor needs to be adjusted carefully, since the DC voltage at the output node shifts significantly with a change of just 10 mV at the base. Third, the signals at the two output nodes are in phase, and the distortion is very low even with a large common-mode input signal.

> To measure the common-mode gain in the lab, the amplitude of the input signal needs to be large because the common-mode gain is very low. On the other hand, distortion of the output signal is not a concern, since the output resistance of the tail transistor provides very deep negative feedback.

Figure 4.9 illustrates the same DA circuit with a single-ended differential-mode input configuration, along with the corresponding simulated Bode plot. While this setup offers a convenient way to simulate differential-mode behavior, it is not entirely rigorous. In practice, the signals observed at the collector nodes contain both differential-mode and common-mode components. On the left branch, the common-mode and differential-mode signals are in phase, whereas on the right branch, they are 180° out of phase. However, this phase difference does not significantly affect the simulation results, because the common-mode component at the output nodes is extremely weak due to the inherent common-mode rejection of the circuit.

The Bode plot depicted in figure 4.9(b) indicates that the differential-mode voltage gain is 31.2 dB, which is consistent with the results shown in figure 4.9(a), $A_d = \pm\, 36.3$ V/V. This gain reflects the amplifier's response to input differences while rejecting common-mode disturbances. Additionally, the plot shows that the

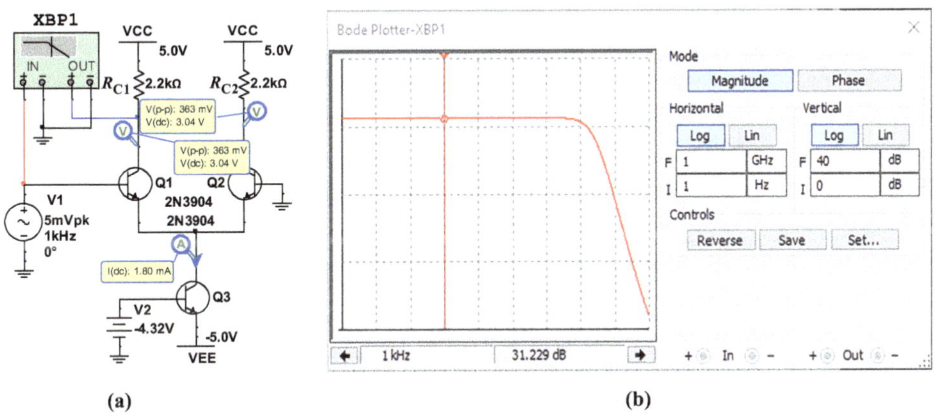

Figure 4.9. (a) DA circuit with a differential-mode input, (b) Bode plot.

gain remains flat in the low-frequency range, indicating that the amplifier has no low-frequency cutoff. This behavior arises from the absence of coupling or bypass capacitors—an intrinsic characteristic of many DA designs, especially those intended for ICs. The flat low-frequency gain underscores the circuit's ability to amplify DC or very low-frequency signals reliably, making it suitable for precision analog applications.

4.3 Simple current mirrors

As discussed in the previous chapter, biasing in discrete amplifier circuits is typically achieved using voltage sources. However, in integrated amplifier circuits, current sources are preferred for biasing due to their superior performance in terms of stability, noise immunity, and ease of integration—though voltage sources are still required in certain parts of the design. To support multistage amplifier architectures, multiple precise current sources must be provided, and this is efficiently accomplished using current mirror circuits. In essence, a current mirror replicates a reference current through one or more additional branches, effectively creating multiple current sources from a single reference. The reference current is usually established by a resistor and a bias voltage or a bandgap reference. By adjusting the size of the transistors in the mirror branches, the mirrored currents can be scaled up or down relative to the reference current. This scalability and simplicity make current mirrors indispensable in analog ICs, where compactness, matching, and thermal tracking are crucial.

Figure 4.10 shows basic current mirror circuits implemented using BJTs and MOSFETs. We will analyze the BJT circuit in more detail, and the MOSFET current mirror can be analyzed in a similar way. The left-hand side of the current mirror is the reference circuit, with transistor Q1 in the diode-connected configuration, resulting in a collector–emitter voltage of about 0.7 V. With this approximation, the reference current can be selected by choosing R_{ref} appropriately:

$$I_{\text{ref}} = \frac{V_{\text{CC}} - 0.7}{R_{\text{ref}}}. \tag{4.10}$$

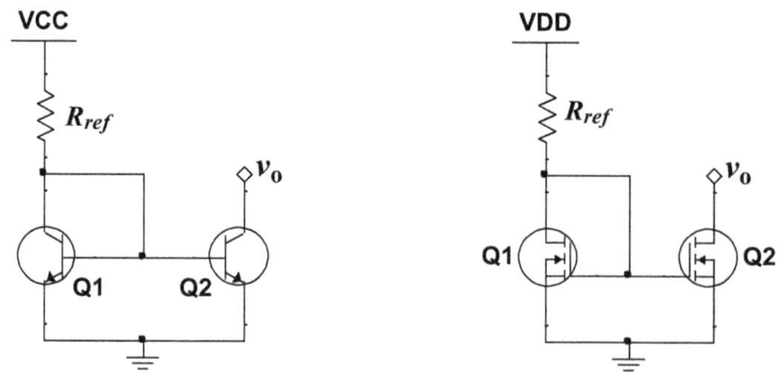

Figure 4.10. Basic current mirror with (a) BJTs and (b) MOSFETs.

For the MOSFET current mirror shown in figure 4.10(b), the drain currents of the two transistors should be identical, provided that the devices are well matched and the drain-to-source voltages are equal. This symmetry ensures accurate current replication, making MOSFET current mirrors highly suitable for ICs. In contrast, the BJT current mirror shown in figure 4.10(a) presents additional complexity due to the presence of a base current, which introduces asymmetry into the current distribution. Specifically, the current flowing through the reference resistor must supply not only the collector current of transistor Q_1 but also the combined base currents of both Q_1 and Q_2. As a result, the reference current I_{ref} splits as follows: $I_{\text{ref}} = I_{\text{C1}} + 2I_{\text{B}}$. Assuming the collector voltages are matched, the relationship between the collector current of Q_2 and the reference current can be found:

$$I_{\text{C2}} = \frac{\beta}{\beta + 2} I_{\text{ref}}. \tag{4.11}$$

Since the current gain β of a typical BJT is usually greater than 100, the resulting mismatch between the reference current and the mirrored current due to the base current is relatively small and often acceptable in practical designs. However, another source of inaccuracy arises from the finite output resistance of the transistor. When the voltage at the output node V_{O} changes, the collector current of Q_2 varies slightly due to the Early effect. This variation can be modeled by incorporating the transistor's output resistance r_{o}, leading to the following expression for the output current:

$$I_{\text{O}} = I_{\text{C2}} + \frac{V_{\text{O}} - 0.7}{r_{\text{o}}}. \tag{4.12}$$

This equation illustrates that the output current is not perfectly constant but exhibits some dependence on the output voltage, reflecting the non-ideal behavior of the current source. For a MOSFET current mirror, a similar relationship holds. However, instead of $V_{\text{BE}} = 0.7$ V, the second term involves the gate–source voltage V_{GS}, which depends on the operating point of the reference transistor.

Figure 4.11 shows the simulation results for a simple BJT current mirror, where the collector current of Q2 is set to 1 mA. As we know, this current is mainly

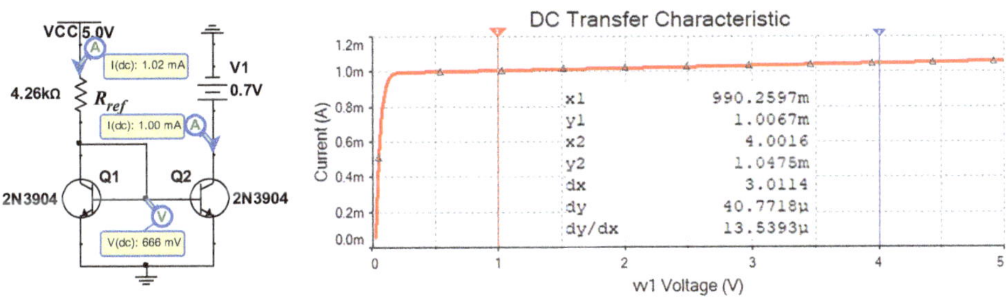

(a) (b)

Figure 4.11. (a) Basic current mirror circuit, (b) simulation results obtained by applying a DC sweep to the voltage source.

controlled by the base–emitter voltage, which is an exponential function. Therefore, the value of the reference resistor R_{ref} needs to be designed using the trial-and-error method. On the other hand, the collector current is also weakly dependent on the collector–emitter voltage in the active mode; thus, a DC voltage source $V1$ is connected to the collector node of Q2 for simulation.

Figure 4.11(b) displays the simulated I–V characteristics of the current mirror when the voltage V_1 at the output node is swept from 0 to 5 V. The curve can be divided into two distinct regions. On the left-hand side, when ($V1 < 0.2$ V), the circuit operates in the saturation region, where the output transistor is not fully active, and the collector current increases sharply with voltage. In contrast, when ($V1 > 0.2$ V), the transistor enters the active mode. In this region, the collector current becomes nearly constant with respect to the output voltage, and the curve flattens out—demonstrating the current source behavior of the mirror.

The small but nonzero slope of the curve in the active region reflects the finite output resistance of the current mirror. This slope can be measured using two cursors on the I–V plot, with the differential conductance ΔI/ΔV indicated by a parameter at the bottom of the data table in figure 4.11(b). The output resistance r_o of the current mirror can then be calculated as the reciprocal of this slope: $R_{om} = 1/13.5\,\mu \approx 74.1$ (kΩ).

This result confirms that the output resistance of a simple current mirror is relatively low, limiting its ability to act as an ideal current source. To overcome this limitation, advanced current mirror configurations are needed.

A simple way to boost the output resistance is by adding an emitter degeneration resistor below the emitter of Q2, which is called a Widlar current mirror. However, to maintain symmetry, the same resistor is also placed below the emitter of Q1. The resulting circuit is shown in figure 4.12(a). The reference current can then be easily estimated:

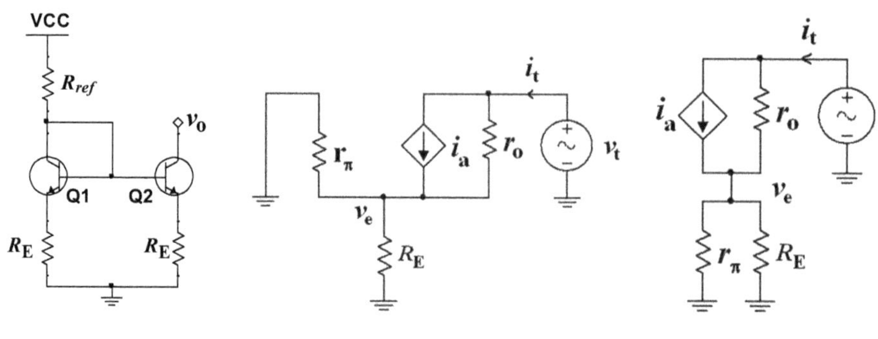

Figure 4.12. (a) Widlar current mirror circuit, (b, c) small-signal circuits.

$$I_{ref} = \frac{V_{CC} - 0.7}{R_{ref} + R_E}. \tag{4.13}$$

We can now determine the output resistance through circuit analysis, using the small-signal circuit shown in figure 4.12(b). As an approximation, the base voltage of these two transistors can be considered fixed by the reference circuit on the left, so it is an AC ground in this circuit. From the perspective of the emitter node, r_π and R_E are in parallel, so this circuit is transformed into the alternative format depicted in figure 4.12(c).

According to Thévenin's theorem, the output resistance of a circuit can be determined using two equivalent approaches: (1) apply a test voltage and measure the resulting current, or (2) apply a test current and measure the resulting voltage. In this analysis, we adopt the first approach—applying a small test voltage v_t at the output node and measuring the corresponding current i_t flowing into the circuit. This method is particularly convenient for small-signal analysis and simulation-based verification. With the nodal voltage v_e as an independent variable, a KCL equation can be set up:

$$i_t = i_a + \frac{v_t - v_e}{r_o} = \frac{v_e}{r_\pi} + \frac{v_e}{R_E}, \tag{4.14}$$

where $i_a = g_m v_\pi = -g_m v_e$. The denominators make the derivation process cumbersome, so each resistance can be replaced with a conductance, and then the relationship between v_e and v_t can be derived:

$$g_o v_t = (g_m + g_o + g_\pi + g_E)v_e \rightarrow v_e = \frac{v_t}{1 + (g_m + g_\pi + g_E)r_o} \approx \frac{v_t}{(g_m + g_\pi + g_E)r_o} \tag{4.15}$$

Plug in this expression for v_e into equation (4.14), and the expression for the current response can be found. Keep in mind that $(g_\pi + g_E)(r_\pi \| R_E) = 1$:

$$i_t = \frac{v_e}{r_\pi} + \frac{v_e}{R_E} = \frac{v_e}{r_\pi \| R_E} = \frac{v_t}{r_o} \frac{1}{1 + g_m(r_\pi \| R_E)}. \tag{4.16}$$

Finally, the output resistance can be derived:

$$R_{om} = \frac{v_t}{i_t} = r_o[1 + g_m(r_\pi \| R_E)]. \tag{4.17}$$

If r_π and R_E are close, the output resistance is approximately equal to $0.5\beta r_o$. For a Widlar current mirror implemented with MOSFETs, this formula can be revised such that $r_\pi = \infty$ and $r_\pi \| R_E = R_E$, making the expression for the output resistance simpler:

$$R_{om} = r_o(1 + g_m R_E). \tag{4.18}$$

The Widlar current mirror circuit with MOSFETs can also be analyzed from the perspective of a feedback system, and this factor $(1 + g_m R_E)$ is called the *amount of feedback*.

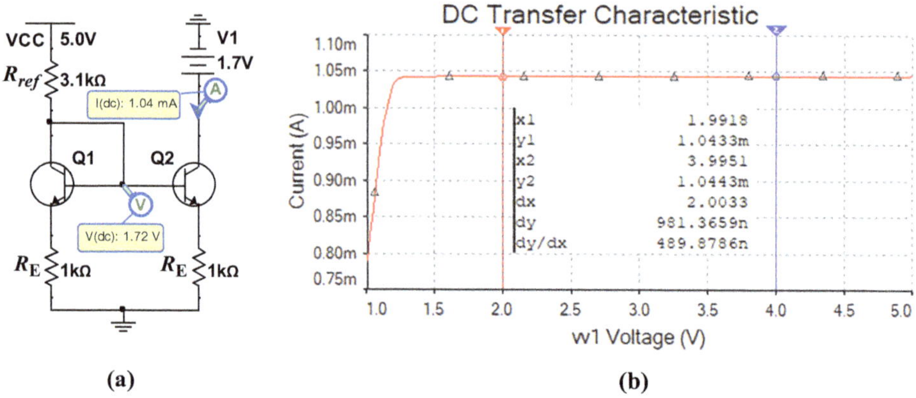

Figure 4.13. (a) Widlar current mirror circuit, (b) simulation results obtained by applying a DC sweep to the voltage source.

Figure 4.13 shows the circuit and the simulation results of a Widlar BJT current mirror. The reference resistor was designed using a trial-and-error method to make the collector current of Q2 close to 1 mA. With a 1 kΩ resistor placed under the emitter, the DC voltage at the base and collector of Q1 is 1.72 V. Figure 4.13(b) illustrates the simulated I–V curve when the voltage of *V*1 is swept from 1 to 5 V; the slope of the curve (dy/dx) is approximately 4.90×10^{-7}. This allows us to determine the output resistance: $R_{om} \approx 2.04\ (\text{M}\Omega)$. Compared to the result for a simple current mirror in figure 4.11, the linear section of this I–V curve looks just like a horizontal line, and the improvement in the output resistance is substantial.

Robert John Widlar (November 30, 1937–February 27, 1991) was an American electronics engineer and a designer of ICs. Widlar invented the basic building blocks of ICs including the Widlar current source, the Widlar bandgap voltage reference and the Widlar output stage. From 1964 to 1970, together with David Talbert, Widlar created the first mass-produced operational amplifier ICs (μA702, μA709), some of the earliest integrated voltage regulator ICs (LM100 and LM105), and the first operational amplifiers employing a single capacitor for frequency compensation (LM101).

4.4 Advanced current mirrors

In ICs, resistors are much more expensive than transistors because they occupy significantly larger chip areas. Therefore, it is more economical to replace resistors with transistors, often resulting in better performance as well. In a cascode current mirror, the emitter degeneration resistor (R_E) on the right branch is replaced by a transistor with an output resistance (r_o) that is usually much larger than R_E. Figure 4.14 shows cascode circuits implemented using BJTs and MOSFETs. The output resistances can be determined using equations (4.17) or (4.18):

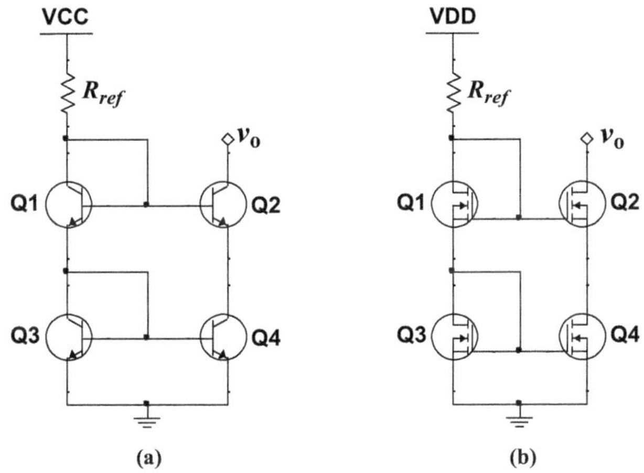

Figure 4.14. Cascode current mirrors: (a) BJT circuit, (b) MOSFET circuit.

$$\begin{aligned} &\text{BJT: } R_{om} \approx r_{o2}[1 + g_{m2}(r_{\pi 2} \| r_{o4})] \approx r_{o2}(1 + g_{m2} r_{\pi 2}) \approx \beta_2 r_{o2} \\ &\text{MOS: } R_{om} \approx r_{o2}(1 + g_{m2} r_{o4}) \approx g_{m2} r_{o2} r_{o4}. \end{aligned} \tag{4.19}$$

The cascode current mirror is a popular configuration with high output resistance, but it has one drawback: the voltage floor at the output node (v_o) is quite high. In modern ICs, VCC or VDD is relatively low and continues to decrease. To provide more headroom, the minimum voltage at the output node needs to be reduced. Although the diode connection ensures that the transistor operates in active mode, it is overly conservative. For instance, the requirement for BJTs to remain in active mode is $V_{CE} > 0.2$ V or lower, but the diode connection sets V_{CE} at 0.7 V. Therefore, breaking the diode connection can lower the voltage floor of these current mirror circuits, but it requires an additional subcircuit to generate the bias voltage at the base or gate.

The complexity of the circuit and the high output resistance make simulating a cascode current mirror quite challenging. Instead of performing a DC sweep of the voltage at the output node, we can use a manual approach as shown in figures 4.15(a) and (b). In this method, two different DC voltages are applied to the output node, and the corresponding currents are simulated. Using this data, the output resistance can be determined from the following formula:

$$R_{om} = \frac{V_2 - V_1}{I_2 - I_1}. \tag{4.20}$$

Using the simulation data shown in figures 4.15(a) and (b), the output resistance is found to be $R_{om} \approx 5.88$ MΩ. Since the current changes very little, the precision of the simulation needs to be increased.

Figure 4.15(c) shows an alternative approach: an AC signal source with a DC offset of 2 V is connected to the output node, and the AC current is obtained from a simulation. The ratio between the amplitudes of the AC voltage and the AC current

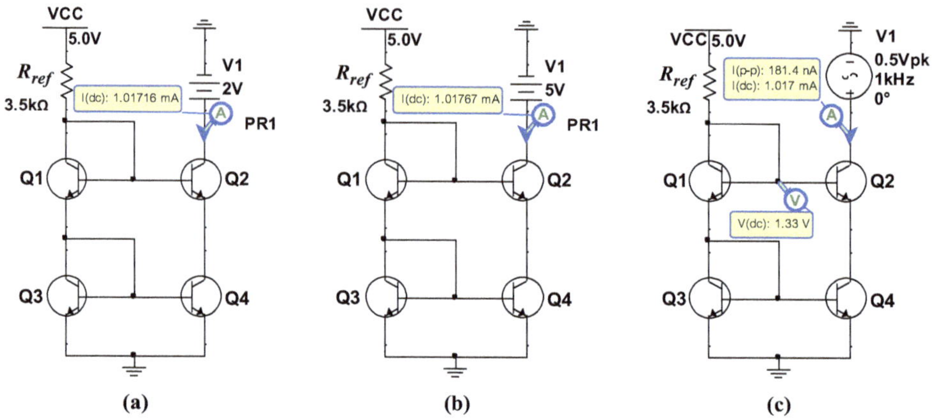

Figure 4.15. Methods used to determine the output resistance: (a, b) DC approach, (c) AC approach.

gives the output resistance: $R_{om} \approx 5.51$ (MΩ). It is important to note the difference between the peak-to-peak amplitude (simulated current) and the conventional amplitude (AC voltage source). The peak-to-peak amplitude needs to be divided by two to convert it into a conventional amplitude. Alternatively, the conventional amplitude can also be converted to a peak-to-peak amplitude, which is more convenient in some situations.

> The output resistance can be further increased by adding another pair of transistors at the bottom with the same connection scheme. However, this approach raises the voltage floor at the output node even higher. Consequently, it is not a viable option in modern ICs.

The Wilson current mirror is another useful configuration, but it is not as popular as the cascode current mirror in modern ICs due to the high voltage floor at the output node. Figure 4.16(a) shows a Wilson current mirror with three transistors; a variation of this circuit is shown in figure 4.16(b), which is more symmetric but has the same characteristics. These MOSFETs can be replaced by BJTs, but then the circuit analysis becomes more challenging due to the base current. To further simplify the analysis, an ideal current source is used at the top of the circuit, which provides the reference current.

As discussed earlier, the output resistance of a network can be determined by applying a test voltage and measuring the response current. However, for the analysis of this circuit, we adopt a different approach: applying a test current and measuring the response voltage. Since the two circuits in figure 4.16 have similar performance, we analyze the simpler one with three transistors. Unlike the cascode current mirror, the Wilson current mirror relies on the feedback loop formed by these transistors.

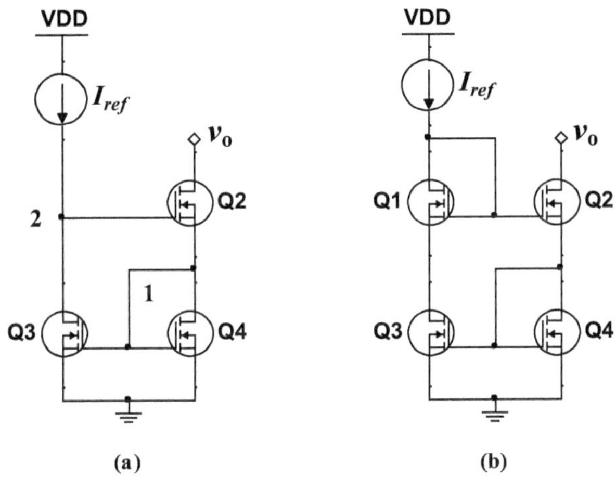

Figure 4.16. Wilson current mirrors: (a) three-transistor circuit, (b) four-transistor circuit.

Analysis of feedback systems often involves a perturbative approach, which can reveal the polarity and sensitivity of various parameters. For example, the current of a MOSFET is controlled by both v_{GS} and v_{DS}, and their influence can be determined using the following equation:

$$\Delta i_D = g_m \Delta v_{GS} + \Delta v_{DS}/r_o = g_m \Delta v_{GS} + g_o \Delta v_{DS}. \tag{4.21}$$

Since $g_m \gg g_o = 1/r_o$, the current is much more sensitive to changes in v_{GS} than changes in v_{DS}. We can imagine a MOSFET as a current source with two control knobs: v_{GS} is the coarse control and v_{DS} is the fine control. Assume the current needs to remain constant; if the coarse control knob (v_{GS}) is adjusted slightly, the fine control knob (v_{DS}) needs to be adjusted significantly in the opposite direction to compensate. This relationship can be described using an expression for voltage gain, which is called the *intrinsic gain*:

$$A_{G-D} = \frac{\Delta v_{DS}}{\Delta v_{GS}} = -\frac{g_m}{g_o} = -g_m r_o. \tag{4.22}$$

For example, $g_m = 20$ mS, $r_o = 50$ kΩ, $A_{G-D} = -1000$ V/V. Therefore, a small change in v_{GS} can cause a significant change in v_{DS}. For example, if $\Delta v_{GS} = 1$ mV, then $\Delta v_{DS} = -1$ V.

Before analyzing the circuit in figure 4.16(a), let us take a closer look at it to identify some useful relationships. First, since the gate current is zero in MOSFETs, there is no 'horizontal current' in this circuit. As a result, the 'vertical currents' remain the same. More specifically, Q2 and Q4 share the same current, and the current of Q3 is constant and equal to the reference current from the top. Second, in addition to the output node, there are only two electrical nodes in this circuit, labeled '1' and '2' in the diagram. Third, the gate and drain of Q4 are tied together in a diode-connected configuration, and the voltage at Q4 is denoted by v_1. This means the drain current of Q4 is solely controlled by v_1, so a simple relationship between the output current i_o and v_1 can be established:

$$\Delta i_o = \Delta i_{D4} = (g_{m4} + g_{o4})\Delta v_1 \approx g_{m4}\Delta v_1 \;\rightarrow\; \Delta v_1 \approx \Delta i_o / g_{m4}. \tag{4.23}$$

On the other hand, v_1 is also the gate voltage of Q3, which is related to v_2 by means of equation (4.22). If we assume that Q3 and Q4 are matched transistors with the same transconductance, the following relationship can be derived:

$$\Delta v_2 \approx -g_{m3} r_{o3} \Delta v_1 \approx -r_{o3} \Delta i_o. \tag{4.24}$$

Compared to the voltage swing at the gate of Q2 (Δv_2), the AC component at the source (Δv_1) is very small. In addition, the current change Δi_o is also very small. Similarly, the corresponding change in current is also negligible. Therefore, equation (4.22) can be applied to Q2:

$$\Delta v_o \approx -g_{m2} r_{o2} \Delta v_2 \approx g_{m2} r_{o2} r_{o3} \Delta i_o \tag{4.25}$$

Finally, the expression for the output resistance can be derived, as follows:

$$R_{om} = \frac{\Delta v_o}{\Delta i_o} \approx g_{m2} r_{o2} r_{o3} \rightarrow g_m r_o^2. \tag{4.26}$$

If the MOSFETs are replaced by BJTs, the derivation process is a little more complex due to the presence of base currents. In addition, the output resistance of a BJT Wilson current mirror is about half that of a cascode current mirror: $R_{om} \approx 0.5\beta_2 r_{o2}$.

Figure 4.17 presents the simulation results for a Wilson current mirror based on BJTs. In the right branch, an AC current source provides the AC stimulus. The DC current component for this source is determined through a trial-and-error method, and it is highly sensitive due to the very high output resistance. This value, specified in the diagram as I_{DC} = 1.004 mA, is slightly lower than the current in the left branch due to the nonzero base current.

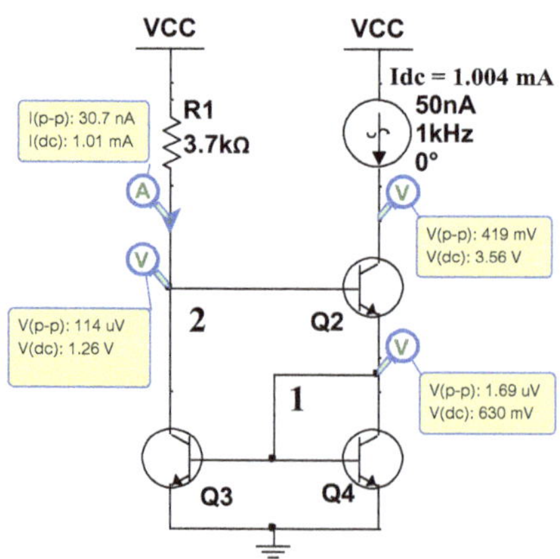

Figure 4.17. Wilson current mirror with three BJTs.

The AC voltages at the three nodes are shown in figure 4.17: at node 1, the amplitude is very low ($V_{1,\text{p-p}} = 1.69\ \mu\text{V}$), which justifies the approximation made in deriving equation (4.25). At node 2, the signal is amplified ($V_{2,\text{p-p}} = 114\ \mu\text{V}$) but the gain is relatively low due to the reference resistor R1 and the input resistance of Q2. At the output node, the signal is amplified significantly ($V_{\text{O,p-p}} = 419\ \text{mV}$). The output resistance can be calculated as follows: $R_{\text{om}} = 4.19\ \text{M}\Omega$.

George R. Wilson was an accomplished IC design engineer. In 1967, he and Barrie Gilbert worked for Tektronix. One day, they challenged each other to find an improved current mirror overnight that would use only three transistors. Wilson won the challenge by designing this current mirror circuit.

4.5 Discrete amplifiers with active loads

Before discussing this new topic, let us briefly review the CE amplifier. The voltage gain of the core amplifier is $A_{\text{VO}} = -g_{\text{m}}(R_{\text{top}} \| r_{\text{o}})$, where R_{top} represents R_{C} for a BJT amplifier and R_{D} for a MOSFET amplifier. Assume that the transconductance g_{m} remains constant. This gain is primarily determined by the value of the resistor R_{top}, since the output resistance of the transistor is usually much higher, i.e. $r_{\text{o}} \gg R_{\text{top}}$. For historic reasons, this resistor R_{top} is called a load resistor. For example, Figure 3.7 in the previous chapter shows a diagram of load-line analysis, where the term 'load line' refers to the I–V characteristics of this resistor.

Here is a question: which device has the highest resistance? The answer is a current source, and its resistance is infinite. Therefore, we can replace the load resistor R_{top} with a current source and see what happens. Figure 4.18(a) shows the circuit with the simulation results, which indicate that the voltage gain is extremely high: $A_{\text{VO}} = -2880\ \text{V/V}$. To verify this result, the Bode plot is also generated, and it agrees reasonably well. Additionally, this result can be compared with the analytic result:

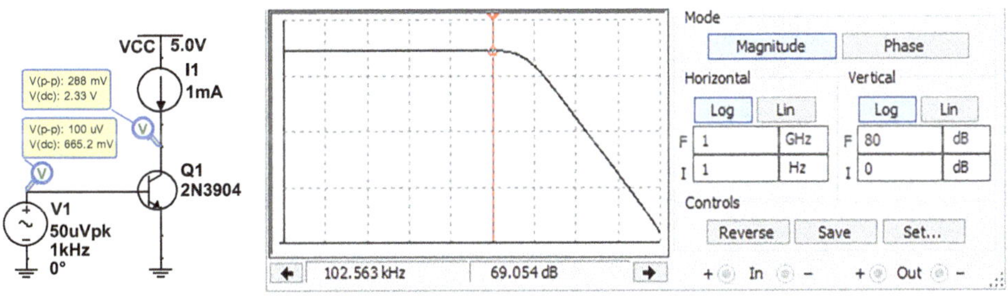

(a) (b)

Figure 4.18. CE amplifier with a current source: (a) circuit, (b) Bode plot.

$$A_{VO} = -g_m r_o = -\frac{V_{An}}{V_T}. \tag{4.27}$$

As discussed earlier, this expression is called intrinsic gain, since it depends solely on the parameters of the transistor. The Early voltage in this equation can be obtained from the device model provided by Multisim: $V_{An} = 74\ V$. At room temperature, the intrinsic gain is −2857 V/V, which agrees quite well with the simulation results.

As discussed in section 3.3, the DC voltage at the output node of this type of amplifier should ideally be centered between V_{CC} and ground to ensure maximum output swing and proper linear operation. In the circuit shown in figure 4.18(a), this is accomplished by carefully adjusting the DC offset of the signal source. As illustrated, this DC voltage is set to $V_{DC} = 665.2$ mV. Due to the high gain and sensitivity of this amplifier, even a tiny deviation—such as a 0.1 mV change—can result in a substantial shift in the collector voltage. This example highlights a classic trade-off in amplifier design: high gain enhances signal amplification but also increases sensitivity to small DC offsets and component mismatches.

The circuit in figure 4.18(a) illustrates the idea that very high gain can be achieved with an active load. However, the current source is impractical for real circuits. As discussed in the previous two sections, a current mirror is a good approximation for a current source. In this simple circuit, a complementary type of transistor functions as the active load. Figure 4.19(a) shows a circuit with a *pnp* BJT as the active load, as well as the simulation results. In contrast to the circuit with a current source as the load, the voltage gain is reduced to: $A_V = -612\ \text{V/V}$.

The design of this circuit is very challenging because the DC bias voltages at the bases of both transistors need to be fine-tuned. Fortunately, the bias voltage of the *npn* BJT in the circuit from figure 4.18(a) can be reused, and then the bias voltage for the *pnp* BJT can be determined using the trial-and-error method. The voltage gains of these two circuits shown in figure 4.19 can be expressed in the same way, as the transconductance remains unchanged if the current is the same:

$$A_{VO} = -g_m(r_{on} \| r_{op}) = -\frac{1}{V_T}\frac{V_{An} \cdot V_{Ap}}{V_{An} + V_{Ap}}. \tag{4.28}$$

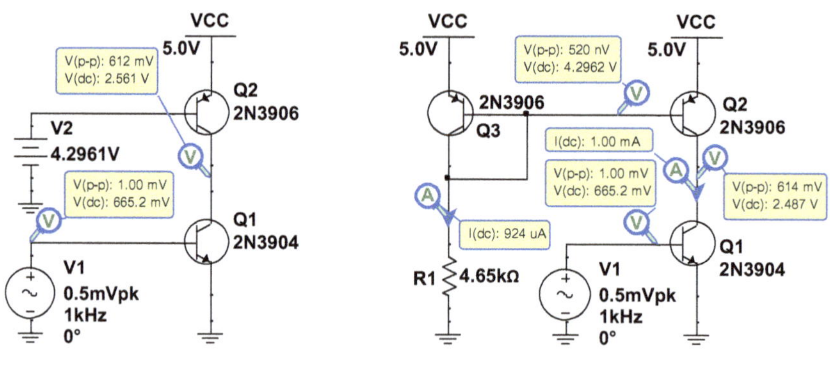

Figure 4.19. CE amplifier with active load, (a) with voltage source, (b) with current mirror.

The Early voltage of this *pnp* BJT listed in the device model is $V_{Ap} = 18.7$ V, which is quite low for a discrete BJT. At room temperature, $V_T = 25.9$ mV, and the calculated voltage gain is −576 V/V. Although this is slightly lower than the simulation result, the difference is not significant.

The circuit shown in figure 4.19(b) is a variation of this amplifier that employs a current mirror as an active load. Although this circuit appears more complex at first glance, it actually simplifies the bias design by eliminating the need to precisely tune the DC bias voltage of the *pnp* BJT. In the first step of the design process, the reference resistor R_1 is chosen to establish the target bias current—approximately 1 mA in this case. In the second step, the DC offset voltage of the signal source is adjusted to set the output node to a voltage near the midpoint between V_{CC} and ground, ensuring a symmetrical output swing. This adjustment can be efficiently carried out using the *DC sweep* simulation tool in Multisim, allowing for a quick search for the optimal bias point.

Compared to the circuit shown in figure 4.19(a), where the base of the *pnp* BJT is connected to a DC voltage, there is a small AC voltage component at the bases of the two *pnp* BJTs in the current mirror. However, the amplitude of this AC component is very small due to the high output impedance of the current mirror, and thus it has minimal impact on the amplifier's gain.

> In ICs, operating currents are typically very low to reduce power consumption and heat. As a result, the transconductance of transistors is also low, which would normally limit voltage gain. To overcome this, active loads are used in place of passive resistors. These active components offer much higher output resistance, significantly boosting gain even at low current levels. This approach is impractical in discrete circuits due to bias sensitivity, but in ICs, precise matching and controlled fabrication make active loads highly effective for building compact, low-power, and high-gain analog amplifier stages.

4.6 Differential amplifiers with active loads

The performance of DAs can be improved by replacing the traditional resistive loads with transistors or current mirrors. This configuration offers higher gain and an improved CMRR. Figure 4.20 shows two DA circuits designed using BJTs and MOSFETs, respectively. The principle is straightforward: the two *p*-type transistors at the top provide an output resistance (r_{op}) that is much higher than that of the replaced resistors, thereby increasing the gain of the DA. However, as discussed in the previous section, the challenge lies in the DC bias voltage. In these two circuits, an auxiliary bias circuit is needed to generate the DC bias voltages V_{B0} and V_{B1} so that the currents are balanced and the DC voltage at the output node is at the midpoint.

The AC analysis of this circuit is straightforward, and the voltage gain in each branch is the same as that of a CE/CS amplifier:

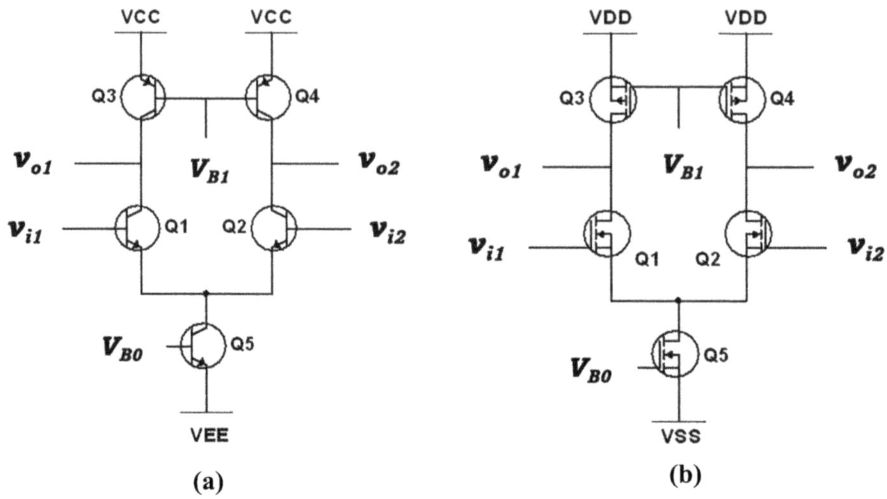

Figure 4.20. DAs with differential outputs: (a) BJT circuit, (b) MOSFET circuit.

$$A_{branch} = \frac{v_{o1}}{v_{i1}} = \frac{v_{o2}}{v_{i2}} = -g_m(r_{on}||r_{op}). \quad (4.29)$$

When written in the format of a differential input: $v_{i1} = \frac{1}{2}v_d$ and $v_{i2} = -\frac{1}{2}v_d$, the gain is reduced by half:

$$A_d = \frac{v_o}{v_d} = \pm\frac{1}{2}g_m(r_{on}||r_{op}). \quad (4.30)$$

If the output is taken from the difference of the two output nodes, the fully differential voltage gain is doubled:

$$A_{d,d} = \frac{v_{o2} - v_{o1}}{v_d} = g_m(r_{on}||r_{op}) \quad (4.31)$$

In many situations, only a single-ended output is needed. Figure 4.21 shows two DA circuits with an active load; the BJT and MOSFET versions operate in the same way. Transistor Q3 is in a diode-connected configuration, eliminating the need for an intricate bias circuit. The output is available only from the right-hand side, and the voltage gain is doubled compared to equation (4.30):

$$A_d = \frac{v_o}{v_d} = g_{mn}(r_{o2}||r_{o4}) = g_{mn}(r_{on}||r_{op}). \quad (4.32)$$

Before delving into the details, let us analyze the circuit intuitively. Suppose a positive differential input is applied to the bases/gates of Q1 and Q2; the voltage on the left-hand side is then higher than that on the right-hand side. More current is steered to the left branch, which causes the base/gate voltage of Q3 to drop. In other words, $|v_{BE}|$ or $|v_{GS}|$ is increased for *p*-type transistors.

Unlike the consistent changes on the left branch, there is a conflict on the right branch. The negative input voltage at the base/gate of Q2 reduces the current, but

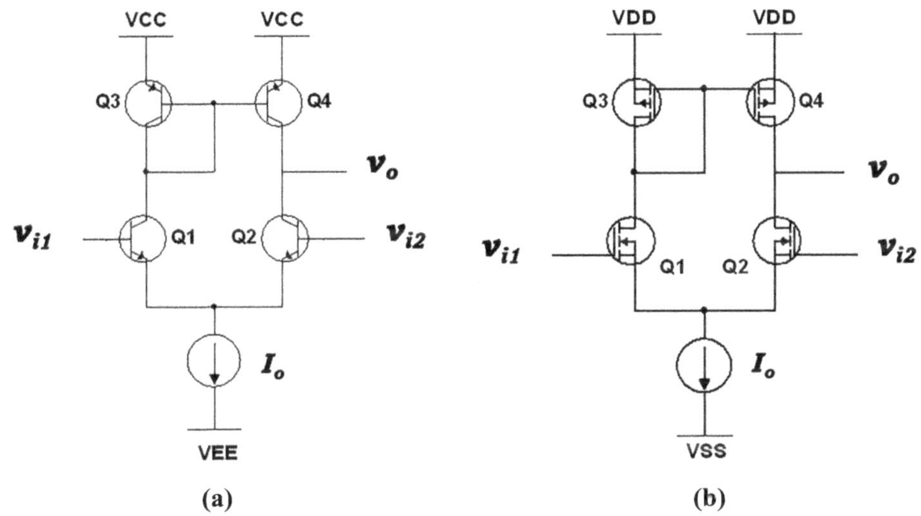

Figure 4.21. DAs with single-ended outputs: (a) BJT circuit, (b) MOSFET circuit.

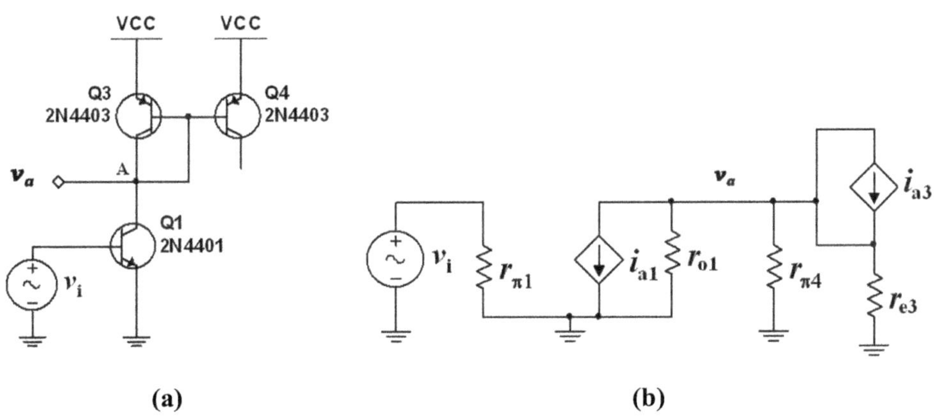

Figure 4.22. (a) Partial circuit of DA, (b) small-signal circuit.

the current mirror at the top tries to increase the current through Q4. This conflict is resolved by adjusting the voltage at the output node, which is the collector/drain node of Q2 and Q4. As discussed previously, a transistor can be modeled as a current source with two control knobs: coarse control at the base/gate and fine control at the collector/drain. To balance the coarse control with the fine control, significant changes are required, leading to a high voltage gain. In addition, this 'double strike' effect from both the top and the bottom doubles the gain compared to equation (4.30).

To simplify the analysis, the transistor Q2 can be cut out first, and the remaining circuit is shown in figure 4.22(a). In the first step, we need to determine the response of the base voltage of the *pnp* BJTs to the input signal. This node, indicated by 'A' in the diagram, is connected to four transistor terminals: the base nodes of the two *pnp* transistors and the collectors of the two transistors in the left branch. Figure 4.22(b) shows the small-signal circuit: the hybrid-π model is used for Q1 and Q4, while the T-model is used for Q3. In this circuit, the diode-connected transistor Q3 is reduced to its emitter resistance r_{e3}, and the transistor Q4 is reduced to the input resistance $r_{\pi 4}$. These two resistors are in parallel with the output resistance of Q1 (r_{o1}), and the total resistance of these three resistors is $R_L = r_{o1} \| r_{\pi 4} \| r_{e3} \approx r_{e3}$, since $r_{o1} \gg r_{\pi 4} \gg r_{e3}$. The following relationship can now be obtained:

$$v_a = -g_{m1} R_L v_i \approx -g_{m1} r_{e3} v_i \approx -\left(g_{m1}/g_{m3}\right) v_i = -\left(g_{mn}/g_{mp}\right) v_i. \quad (4.33)$$

Although this equation is derived from the circuit with BJTs, it is also valid for the same circuit with MOSFETs, since $r_{e3} \rightarrow r_{gs3} = 1/g_{m3}$ for MOSFETs. In fact, it is easier for MOSFETs, since $r_{\pi 4} \rightarrow \infty$.

In the second step, we focus on the circuit in the right branch containing transistors Q2 and Q4. These two transistors can be replaced with their hybrid-π small-signal models, as shown in figure 4.23(a). For the transistor Q4, the input signal is $v_a = -0.5\left(g_{mn}/g_{mp}\right) v_d$, and then the current is $i_{a4} = -0.5 g_{m4} \cdot (g_{m1}/g_{m3}) v_d = -0.5 g_{mn} v_d$, which is the same as the current i_{a2}. Therefore, these two current sources can be combined, as well as the two output resistors, resulting in the simplified circuit shown in figure 4.23(b). Finally, the output signal can be derived from this circuit: $v_o = -2 i_a (r_{o2} \| r_{o4}) v_d = g_{mn} (r_{o2} \| r_{o4}) v_d$.

Figure 4.24(a) shows the simulation results for this DA circuit, with an ideal current source at the bottom that can be replaced with a current mirror. First, the voltage gain can be calculated from the displayed result: $A_V \approx 1993$ V/V. With a 1 mA DC current in each branch, $g_m \approx 38.6$ mS, $r_{on} = 90.7$ kΩ, $r_{op} = 115.7$ kΩ, and $A_V \approx g_m\left(r_{on} \| r_{op}\right) \approx 1963\ (\mathrm{V/V})$, which agrees with the simulation results quite well. Second, the signal on the left side is very weak because the diode-connected transistor Q3 provides a low input resistance. Third, the DC current in the right branch is slightly less than half of the total current, since the current in the left branch includes the two base currents. If the BJTs are replaced by MOSFETs, this imbalance in current can be removed. Figure 4.24(b) shows the circuit with single-ended input, and the simulation results are essentially the same, since the common-mode signal is rejected.

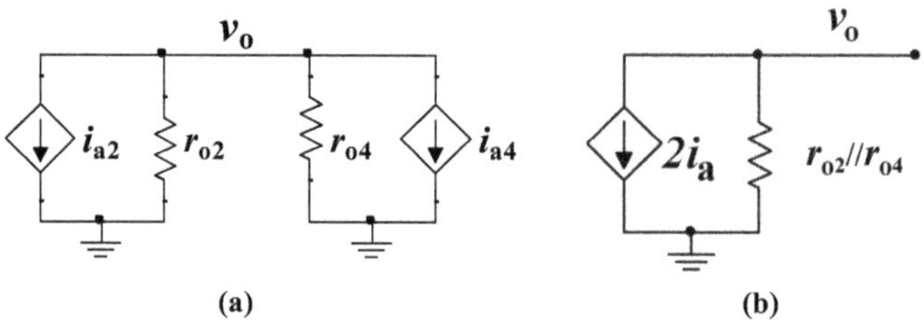

Figure 4.23. (a) Small-signal circuit for Q2 and Q4, (b) an equivalent circuit.

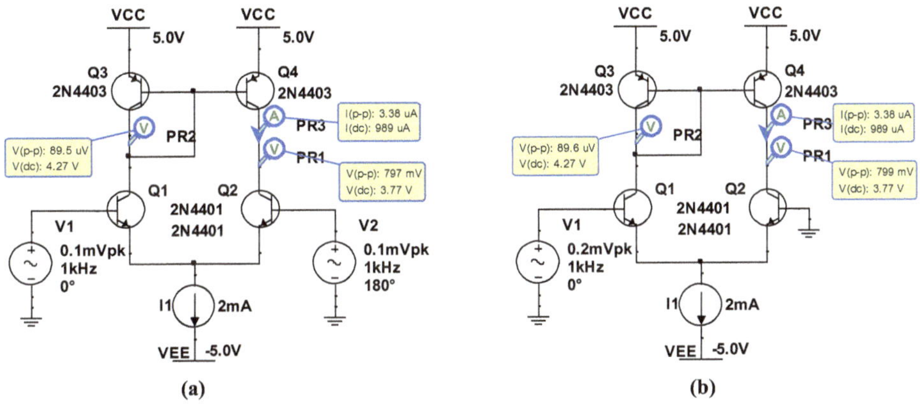

Figure 4.24. Simulations of DAs with active loads: (a) differential input, (b) single-ended input.

> In ICs, the output resistance of MOSFETs with a short channel is much lower than that of discrete transistors, and the current is also much weaker. Therefore, the voltage gain is not very high. To achieve higher gain, more advanced circuit configurations are needed. There are two different approaches: vertical and lateral. The vertical approach engages cascode configurations, which can enhance the output resistance as discussed in the previous section. The lateral approach refers to multistage amplifiers, which are widely used in Op-Amp circuits. These topics are discussed in the following sections.

4.7 Cascode amplifiers

For amplifiers with discrete transistors, the active load configuration provides ample voltage gain. However, this is not the case for ICs, where the short-channel effect significantly reduces the Early voltage of MOSFETs. For example, when the channel length is larger than 1 μm, the Early voltage is usually higher than 50 V. However, when the channel length is reduced to 28 nm, which is the shortest value for a planar MOSFET, the Early voltage drops to around 1 V. Unfortunately, Multisim does not offer good device models for MOSFETs, so we must use BJT circuits to demonstrate the simulation results. To emulate the short-channel effect of MOSFETs, the Early voltages in the BJT device models need to be revised.

First, select a transistor and then double-click it. In the pop-up window shown in figure 4.25(a), select the tab 'Value' at the top and then click the 'Edit model' button at the bottom. The model parameters are displayed in a new window, as shown in figure 4.25(b). The parameter 'Forward Early Voltage' is abbreviated as VAF, and the default value for the 2N4401 transistor is 90.7 V, which is reduced to 5 V in this diagram. Assuming the collector current is 1 mA, the default value of the output resistance is $r_o = 90.7\ \text{k}\Omega$, and its revised value becomes $r_o = 5\ \text{k}\Omega$.

When the output resistance (r_o) is reduced significantly, circuit analysis becomes more complicated, since some approximations taken for granted in the past are no

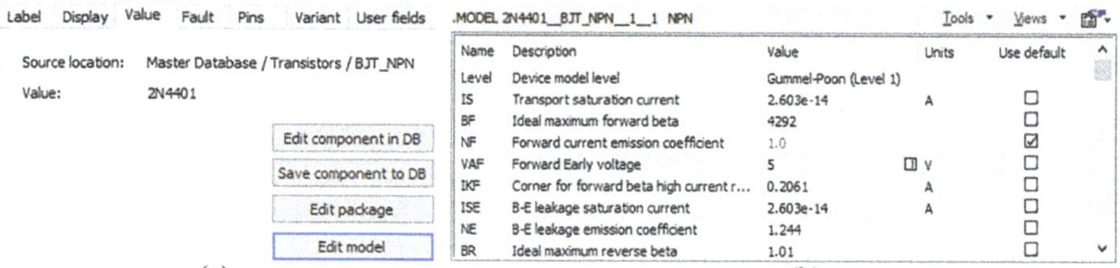

(a) (b)

Figure 4.25. Device model of a BJT: (a) pop-up window, (b) Early voltage parameter—VAF.

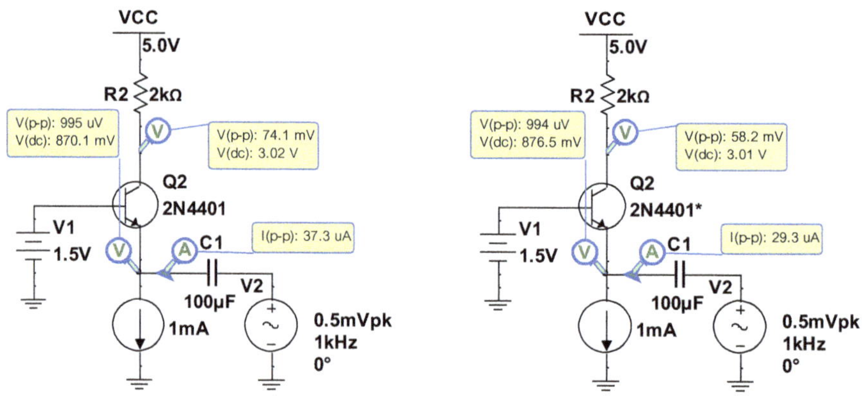

(a) (b)

Figure 4.26. Common-base amplifiers: (a) with $V_A = 90.7$ V, (b) with $V_A = 5$ V.

longer valid. For example, figure 4.26 shows the simulation results for two common-base (CB) amplifiers that differ only in the Early voltage of the BJT. In the circuit on the right, this value is reduced from 90.7 to 5 V. To indicate that the device model has been revised, a star symbol appears after the transistor model number (2N4401*). With the simulation results shown in the circuits, the input resistance at the emitter node of the transistor can be calculated as the ratio between the AC voltage and the AC current at the emitter node: $R_{\text{in,a}} \approx 26.7\ \Omega$ and $R_{\text{in,b}} \approx 33.9\ \Omega$. As discussed in the previous chapter, the input resistance for a CB amplifier can be found easily using the T-model: $R_{\text{in}} \approx r_e \approx 25.9\ \Omega$, and this approximation is valid when $r_o \gg R_2$. However, this condition is not met when the Early voltage is reduced to 5 V and $r_o = 5\ \text{k}\Omega$.

To derive the input resistance of this CB amplifier, the hybrid-π model is needed. The procedure is straightforward: apply a test voltage signal at the emitter and measure the resulting current. The ratio of these two values gives the input resistance:

$$R_{\text{in}} = \frac{r_o + R_{\text{LD}}}{1 + g_m r_o + (r_o + R_{\text{LD}})/r_\pi} \approx \frac{r_o + R_{\text{LD}}}{1 + g_m r_o}. \tag{4.34}$$

In this equation, R_{LD} stands for the load resistor above the transistor. If the condition $r_o \gg R_{LD}$ is met, this equation reverts to its approximated format: $R_{in} \approx 1/g_m \approx r_e$. We can now apply equation (4.34) to find the input resistance of the circuit in figure 4.26(b): $R_{in,b} \approx 36.1\ \Omega$, which agrees reasonably well with the simulation results. In addition, the approximate expression in equation (4.34) becomes accurate for MOSFETs, since $r_\pi \to \infty$.

An extreme case occurs when R_{LD} is replaced by a current source ($R_{LD} \to \infty$), and the input resistance becomes $R_{in} = r_\pi$, which can be derived from the first expression in equation (4.34). In fact, this result can be obtained directly from the small-signal circuit, since a DC current source becomes an open circuit in AC analysis.

4.7.1 Telescopic cascode amplifiers

For transistors with a low Early voltage, the gain can be boosted using a cascode configuration. To illustrate this concept, consider the circuit shown in figure 4.27(a). The load resistor R1 at the top lowers the voltage gain, but the design and analysis are much simpler. Compared to a CE amplifier, the only difference is the addition of Q2, with its base connected to a DC voltage source. This DC voltage can vary widely without affecting the amplifier's performance. However, there are some limitations. First, the emitter voltage of Q2 is the collector voltage of Q1, which should be high enough to prevent Q1 from entering saturation mode. Second, the emitter voltage of Q2 should not be too high either, since this would limit the swing range of the output signal at its collector node.

As discussed in section 4.4, the cascode configuration can increase the output resistance, thereby enhancing the gain of the amplifier. However, with a resistive load (R_1) at the top, the gain of this circuit changes only slightly, from 37.0 to 37.5 dB. Conversely, if an active load is used at the top, the situation is completely different. Additionally, the presence of Q2 can increase the bandwidth of this

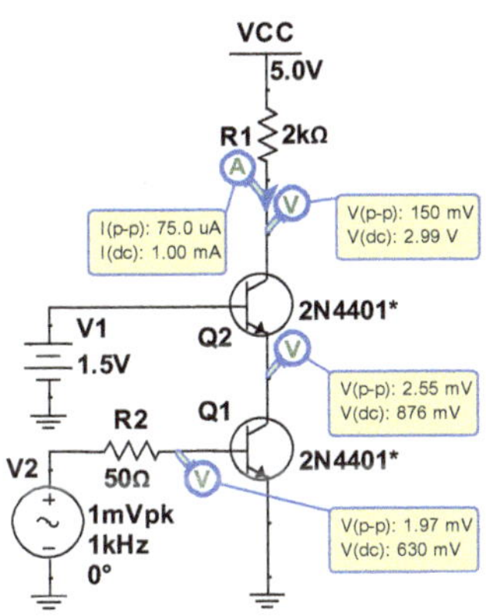

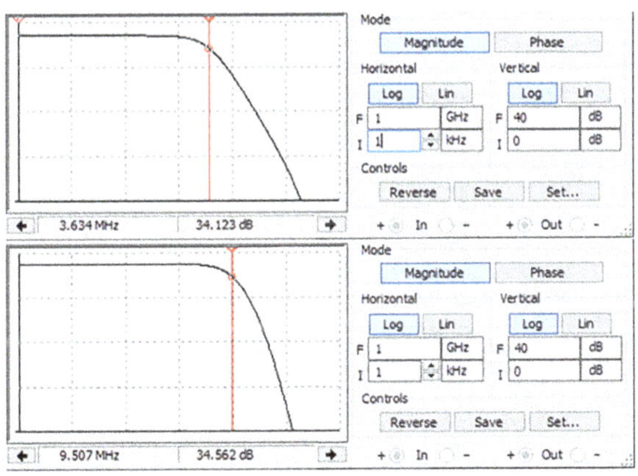

Figure 4.27. Cascode amplifier with a passive load: (a) circuit, (b) Bode plot.

amplifier, as demonstrated in figure 4.27(b). The Bode plot of the CE amplifier without Q2 is at the top, and the one with Q2 is at the bottom. A comparison of these two Bode plots indicates that the cascode amplifier has a higher bandwidth because the Miller effect is eliminated. In other words, there is no direct capacitive coupling between the input and output nodes due to the presence of Q2.

In the next step, the load resistor R_l is replaced by an ideal current source, which is shown in figure 4.28(a). The simulation results indicate that the voltage gain is extremely high: $A_V = -25\,000$ V/V. As we know, higher gain means higher sensitivity, and thus the DC offset voltage of the signal source needs to be adjusted very carefully.

The very high gain of a cascode amplifier can also be explained from a different perspective: it functions as a two-stage amplifier. The first stage is a CE amplifier with Q1, and the gain is $A_{V1} = -g_{m1}(R_{in2} \| r_{o1}) \approx -94.\ 6$ V/V, where R_{in2} stands for the input resistance at the emitter of Q2, which is equal to r_π in this case. The second stage is a CB amplifier with Q2, and its gain is $A_{V2} = g_{m2}(R_{LD} \| r_{o2}) = g_{m2} r_{o2} \approx 264$ V/V.

In the circuit with a resistive load shown in figure 4.27(a), the function of Q2 is to boost the output resistance at the output node: $R_O \approx g_{m2} r_{\pi 2} r_{o2}$. Therefore, the overall voltage gain can be expressed using the parameter of *effective transconductance* (G_m): $A_V = -G_m(R_l \| R_O)$, where $G_m \approx g_{m1}$ in this case. In the small-signal hybrid-π model, the current generated by the controlled source $g_{m1} v_s(t)$ flows primarily into the low-impedance emitter of Q_2, rather than being lost through r_{o1}. This current is then steered through Q2 to the load resistor, maximizing the gain contribution of Q1 while preserving linearity. Thus, the cascode configuration

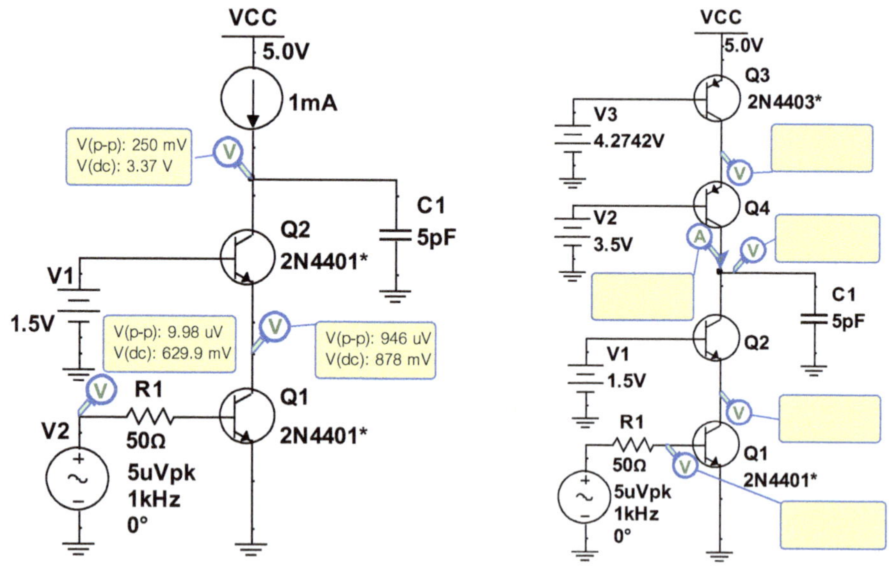

Figure 4.28. Cascode amplifiers: (a) with a current source, (b) with complementary transistors.

improves both gain and output resistance, making it a foundational building block in high-gain analog designs.

Similarly, for the circuit shown in figure 4.28(a), the voltage gain can be expressed in the same way: $A_V = -G_m(R_{LD} \| R_O)$. However, the effective transconductance G_m is quite different from g_{m1}, since the input resistance at the emitter of Q2 is at the same level as r_{o1}. In other words, in the small-signal circuit, a current divider is formed between R_{in2} and r_{o1}, so the 'useful current' going into Q2 is less than the current signal generated by Q1, $g_{m1}v_s(t)$, which comes from the current source in the hybrid-π model. Therefore, the effective transconductance becomes lower:

$$G_m = \frac{r_{o1}}{R_{in2} + r_{o1}} g_{m1}. \tag{4.35}$$

In the circuit shown in figure 4.28(a), a current source is at the top, and the input resistance can be found from equation (4.34): $R_{in2} \approx r_\pi$. The current gain of a 2N4401 transistor is $\beta = 120$, so $R_{in\,2} \approx 3.13$ kΩ. Using equation (4.35), the effective transconductance can be obtained: $G_m \approx 0.615 g_{m1}$.

The current source in figure 4.28(a) can now be replaced with two *pnp* BJTs in a cascode configuration, as shown in figure 4.28(b). The design of this circuit is quite subtle, as the DC voltages at the bases of Q1 and Q3 need to be adjusted very carefully. Although the Early voltages of all the BJTs were lowered to 5 V, the voltage gain remains very high: $A_V = -11\,200$ V/V.

> Since the bias currents in most IC amplifiers are very low, the transistor transconductance is also small, resulting in limited voltage gain. Therefore, the gain of a single-stage amplifier in practical ICs is much lower than the idealized simulation results shown in this section. For example, in telescopic cascode amplifiers using MOSFETs, the voltage gain typically only reaches a few hundred.

4.7.2 Folded cascode amplifiers

The circuit shown in figure 4.28(b) is called a 'telescopic cascode amplifier,' and it works well when V_{CC} is high enough. However, as mentioned before, the voltage of the internal power supplies in modern ICs is around 1 V, making this type of amplifier unsuitable for such conditions. An alternative circuit is the 'folded cascode amplifier' shown in figure 4.29(a), which employs an *npn* BJT, Q1, and a *pnp* BJT, Q2. The Bode plot in figure 4.29(b) indicates that the performance of this amplifier is essentially the same as that of the circuit shown in figure 4.27(a), but the required minimum power supply voltage is lower.

The folded cascode amplifier can be understood from two different perspectives. First, it follows the principle of the two-step approach: a voltage-to-current conversion followed by a current-to-voltage conversion. As shown in the simulation result in figure 4.29(a), the sum of the currents passing through the two transistors is

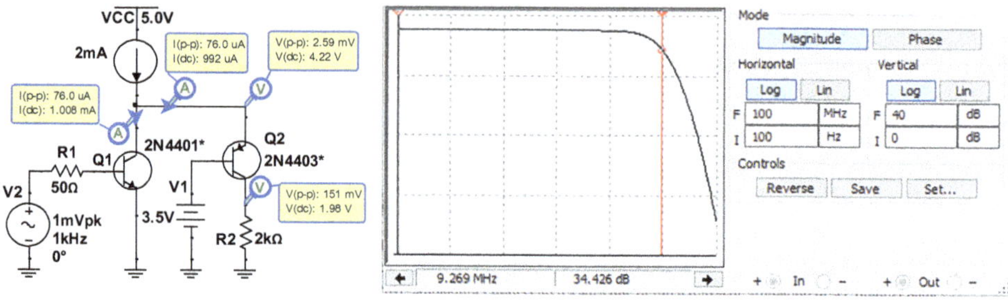

(a) (b)

Figure 4.29. Folded cascode amplifier: (a) circuit, (b) Bode plot.

constant and equal to the current provided by the current source at the top. The simulation results for the AC currents indicate that they have the same amplitude, $I_{\text{p-p}} = 76.0\ \mu\text{A}$, but the opposite signs are not revealed. When the input signal causes a current swing in transistor Q1, the current of transistor Q2 should swing with the same amplitude but in the opposite direction. Ultimately, resistor R_2 converts the current swing into a voltage swing; the gain can be obtained as follows:

$$A_{\text{V}} = -G_{\text{m}}(R_2 \| R_{\text{O2}}) \approx -g_{\text{m1}}(R_2 \| R_{\text{O2}}). \tag{4.36}$$

In this equation, R_{O2} stands for the output resistance of the transistor Q2, similar to that of the cascode amplifier in figure 4.27(a).

An alternative perspective is the two-stage amplifier approach: a CE amplifier with Q1 and a CB amplifier with Q2. The load resistance for Q1 is the input resistance at the emitter of Q2, which can be calculated using equation (4.34): $R_{\text{in}} \approx 36\ \Omega$. Therefore, the gain of the first stage is very low: $A_{\text{V1}} = -1.39$ V/V. As a result, the second stage contributes most of the voltage gain of this amplifier.

4.7.3 Cascode differential amplifiers

The cascode amplifiers discussed so far have a common problem: fine-tuning of DC bias voltages is needed. However, this cannot be implemented in ICs, so these circuits are not practical. Fortunately, no such fine-tuning is needed for DAs, as shown by the example in figure 4.30(a). The four *pnp* BJTs at the top form a cascode current mirror, which offers a very high output resistance at the collector of Q6. Similarly, the four *npn* BJTs at the bottom also offer a very high output resistance at the collector of Q4. The output resistance can be estimated from the cutoff frequency in the Bode plot shown in figure 4.30(b): $R_{\text{O}} = R_{\text{On}} \| R_{\text{Op}} \approx 11.3\ (\text{M}\Omega)$.

Telescopic cascode amplifiers remain a valuable design choice in many applications where high gain and bandwidth are necessary. However, they come with several challenges and limitations. First, due to the stacking of transistors, these amplifiers require a higher supply voltage, which is not feasible in modern ICs with low supply voltages for power efficiency. Second, they have a relatively limited output voltage swing because each transistor in the cascode structure requires a certain amount of headroom to operate properly, reducing the overall available

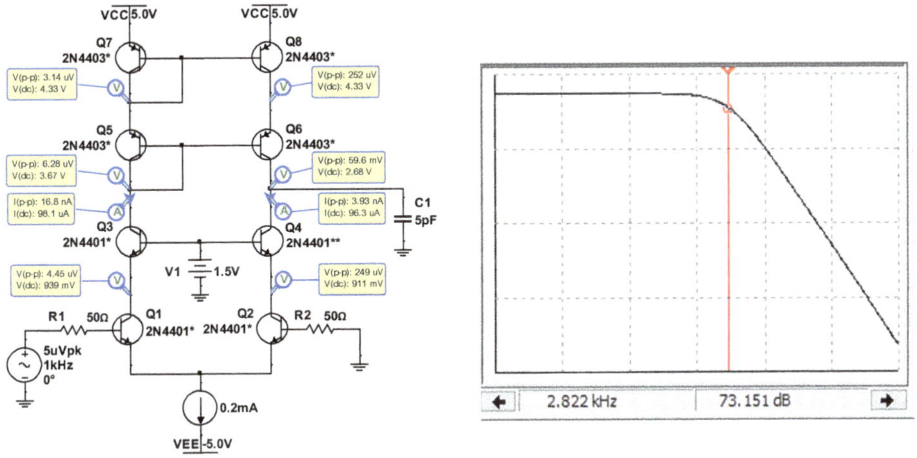

(a) (b)

Figure 4.30. Telescopic cascode amplifier: (a) circuit, (b) Bode plot.

output swing. Third, telescopic cascode amplifiers are sensitive to process variations, affecting the matching and performance of the cascode stages and leading to potential issues with gain, bandwidth, and stability. Fourth, the design complexity is high, resulting in longer design cycles and increased verification and testing requirements.

4.8 Multistage differential amplifiers

Multistage DAs are crucial components in analog circuit design, widely used in various electronic applications such as signal processing, instrumentation, and communication systems. In this type of circuit, multiple amplifier stages are cascaded to achieve higher gain as well as better input and output impedance characteristics. In addition, each stage can be tailored to optimize specific parameters such as gain, bandwidth, or impedance, making the overall amplifier highly adaptable to different applications.

Figure 4.31 shows a simple example of a multistage amplifier, where resistors are used to reduce design and simulation complexity. A drawback of this approach is that the gain is limited without active loads. From the simulation results, the gains can be calculated.

- The first stage is a fully DA, $A_{\mathrm{d,d}} = 62.4$ V/V.
- The second stage is a single-ended amplifier, $A_{\mathrm{d}} = 27.9$ V/V.
- The third stage is an emitter follower with a gain close to unity.

The overall gain is $A_{\mathrm{V}} = 1740$ V/V. At the bottom of this circuit are several current mirrors, and the simulation results indicate that the collector currents from Q1 to Q3 are very close. On the other hand, Q9 is a large transistor, which provides a higher current for the output stage.

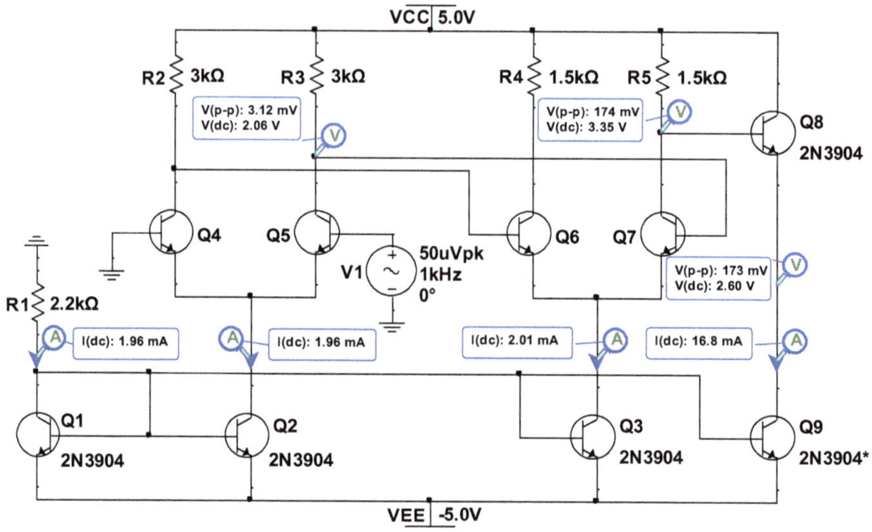

Figure 4.31. Multistage DA circuit.

As the number of stages in an amplifier increases, the frequency response becomes more complex, often introducing stability challenges such as unwanted oscillations, excessive ringing, or degraded transient performance. These problems arise primarily from the cumulative phase shifts introduced by each amplifier stage and the interaction among multiple poles and zeros in the transfer function. As these phase shifts approach or exceed 180°, the negative feedback intended to stabilize the amplifier can effectively become positive feedback, leading to instability. Therefore, ensuring stability is a critical consideration in the design of multistage DAs, especially in high-gain or high-frequency applications.

A key parameter used to evaluate amplifier stability is the *phase margin*, which indicates how close the system is to oscillation. The phase margin is defined as the difference between the actual phase angle and −180° at the gain crossover frequency—the frequency at which the open-loop gain magnitude equals unity (0 dB). For example, consider the amplifier circuit shown in figure 4.31, with its Bode plot depicted in figure 4.32. The magnitude plot at the top shows that the unity-gain frequency occurs at 275 MHz. The corresponding phase angle at this frequency, as shown in the phase plot at the bottom, is −198°. This results in a phase margin of −18°, which indicates instability. A negative phase margin suggests that the amplifier is likely to oscillate unless compensated. In practical designs, a phase margin of 45°–60° is typically desired to ensure robust stability with adequate transient response.

Unstable amplifiers can be converted into stable amplifiers by applying compensation. The simplest approach is to introduce a dominant pole at low frequencies. For example, if a 10 μF capacitor is connected to the output node, introducing a dominant pole at 1.34 kHz, the phase margin becomes 18° as shown in figure 4.33. Unfortunately, the bandwidth of the amplifier is reduced significantly, which is the

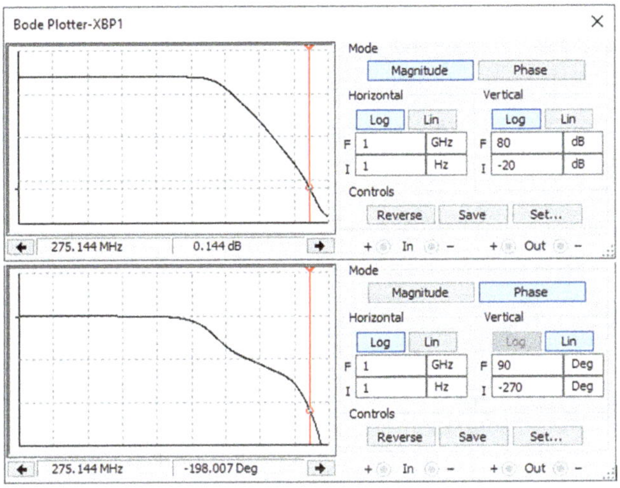

Figure 4.32. Bode plot of multistage DA circuit.

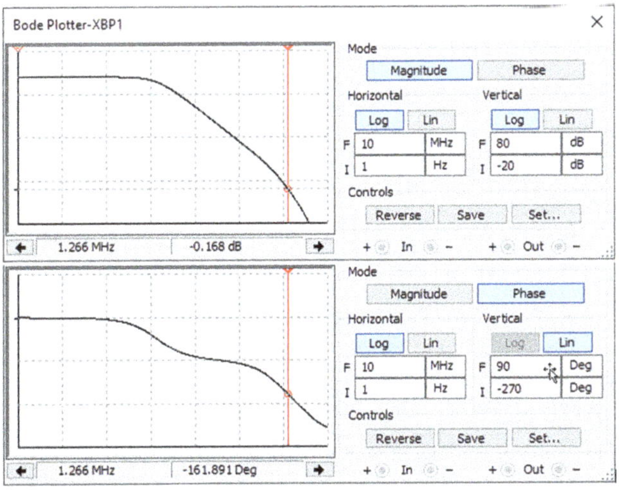

Figure 4.33. Bode plot showing the effect of a 10 μF capacitive load at the output node.

price paid for enhanced stability. In addition, this load capacitor is too large, making it impossible to integrate it into a chip.

The Miller effect can be used to reduce the size of this capacitor, as shown in figure 4.34. This technique involves adding a capacitor between the input and output nodes of a gain stage. The amplified capacitance can create a dominant pole at a lower frequency, thereby improving the phase margin. This method is simple and effective for many designs, but the bandwidth is reduced significantly.

Figure 4.35 shows the Bode plot of the circuit in figure 4.34, with the phase margin improved to 61.4° by introducing two 100 nF capacitors for compensation.

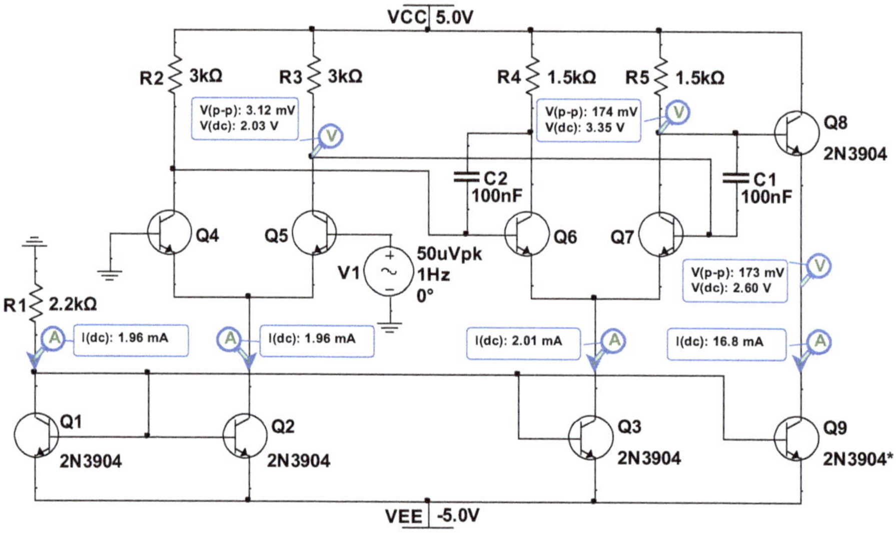

Figure 4.34. Multistage DA circuit with Miller compensation.

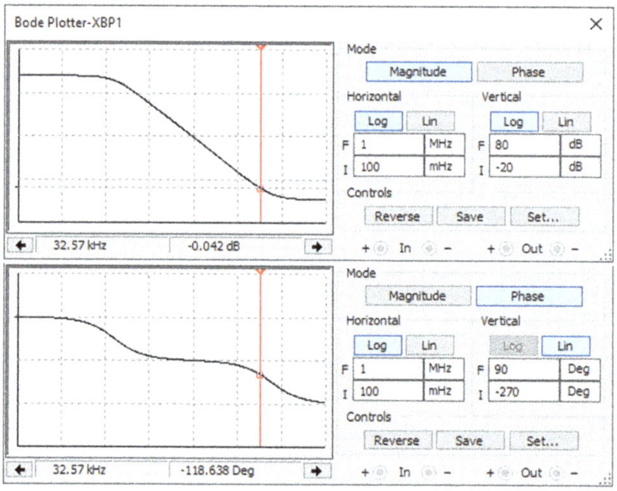

Figure 4.35. Bode plot of the amplifier with Miller compensation.

Unfortunately, the dominant pole is at a very low frequency (16 Hz), so the bandwidth is very limited. Although the compensation capacitance is reduced significantly, it is still too large to be integrated on-chip. However, if active loads are used, the gain becomes much higher, and the value of the capacitors can be further reduced.

DAs are integral components of operational amplifiers (Op-Amps), which are covered in the next chapter. Figure 4.36 shows the internal circuit of the LM741

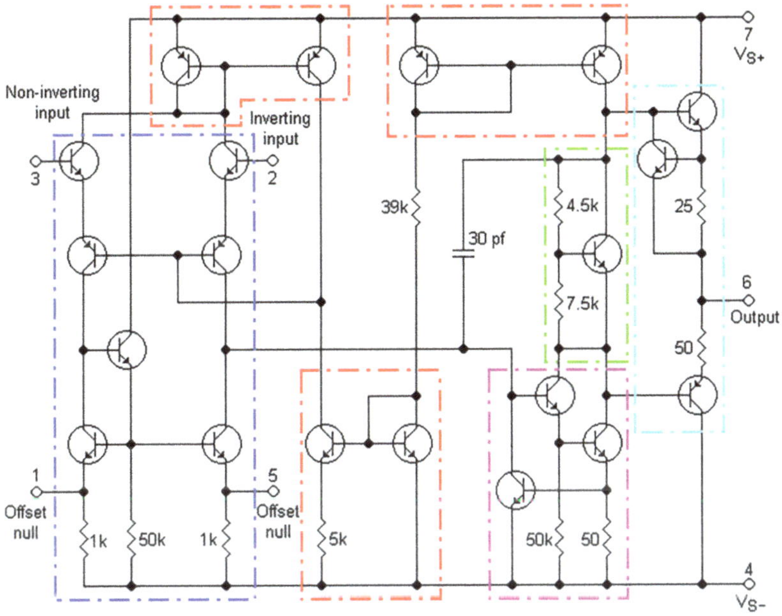

Figure 4.36. Internal circuit of the LM741 operational amplifier. This [OpAmpTransistorLevel Colored] image has been obtained by the author from the Wikimedia website where it was made available by [WvBraun] under a CC BY-SA 3.0 licence. It is included within this book on that basis. It is attributed to [Daniel Braun].

Op-Amp, an early successful model that is still widely used. In the middle of this circuit, the 30 pF capacitor functions as the Miller compensation capacitor. Due to the high gain between the two nodes, this small capacitor is effectively amplified to a very large capacitance, so that the dominant pole of this Op-Amp is around 10 Hz.

> Op-Amps are fundamental building blocks in analog electronic circuits. They are highly versatile and used in a wide range of applications, including signal amplification, filtering, and mathematical operations such as addition, subtraction, integration, and differentiation. Op-Amps typically have a high input impedance, low output impedance, and a very high open-loop gain. These characteristics make them ideal for use in feedback configurations, where they can maintain stable and predictable performance. Op-Amps are crucial in various fields such as audio processing, instrumentation, and control systems.

Chapter 5

Op-Amp circuits

Operational amplifiers, commonly known as Op-Amps, are essential building blocks in analog electronics. Despite their complex internal circuitry—illustrated in figure 4.36 at the end of the previous chapter—their external behavior is very simple. This simplicity, combined with their versatility, makes them suitable for a wide range of applications, including signal amplification, filtering, and mathematical operations. The Op-Amp family includes many variants, each with its own specifications. At its core, an Op-Amp is a high-gain differential amplifier with very high input impedance and typically features a single-ended output. To ensure stable performance, most modern Op-Amps are internally compensated.

In this chapter, we begin by introducing the fundamental characteristics of Op-Amps, including their ideal and practical behaviors, and key parameters such as gain and input impedance. We then explore several basic applications that demonstrate how Op-Amps can be used in real-world circuits. As the name suggests, the primary function of an Op-Amp is signal amplification; therefore, various amplifier configurations—such as inverting, non-inverting, differential, and summing amplifiers—are presented and analyzed. Beyond amplification, Op-Amps play a critical role in active filter design. This chapter covers both first-order and second-order filters, including low-pass, high-pass, and band-pass filters, highlighting their frequency responses and practical design considerations.

We then delve into circuits that utilize negative feedback, emphasizing how feedback improves stability and linearity. Some useful features of simulation tools are also introduced, which provide valuable insights into circuit performance under parameter variations and nonideal conditions, with a focus on using Multisim for sensitivity and distortion analysis. Finally, we examine Op-Amp circuits that employ positive feedback, using the Schmitt trigger as a representative example. This circuit, which demonstrates controlled hysteresis characteristics, is useful in signal conditioning and digital interfacing applications.

doi:10.1088/978-0-7503-5512-4ch5

5.1 Introduction to Op-Amps

Using our knowledge of the differential amplifiers discussed in the previous chapter, we can now identify the pins in the device symbol shown in figure 5.1(a). As we know, complementary biasing is commonly used in differential amplifiers, and thus many Op-Amps also require such a biasing scheme. For example, an Op-Amp might be biased with $V_{S+} = 5$ V and $V_{S-} = -5$ V. However, in some applications, a negative bias voltage may not be available. In such cases, one can select an Op-Amp that operates effectively with a single supply voltage, where the pin of the negative bias voltage is grounded. Additionally, the device's performance is generally not highly sensitive to the exact bias voltages, which can vary over a wide range. As a result, the power supply pins are sometimes omitted from the symbol to simplify the representation.

The circuit shown in figure 5.1(b) provides the basic parameters of the device. Similar to a differential amplifier, the input voltage is defined as the difference between the two input signals: $V_{in} = V^{+} - V^{-}$. These two input nodes are called 'inverting' (V^{-}) and 'non-inverting' (V^{+}) nodes, respectively. When this differential input signal is very weak, the output voltage is proportional to it: $V_{out} = GV_{in}$, where G is the voltage gain, typically higher than 10^5 V/V. In addition, the input resistance R_{in} is very high, usually exceeding 1 MΩ. In MOSFET-based Op-Amps, R_{in} can be replaced by Z_{in}, since the gates of MOSFETs introduce capacitive impedances. The output resistance R_{out} is very low—typically less than 100 Ω. For power Op-Amps, the output resistance is even lower, usually below 1 Ω. Furthermore, the common-mode rejection ratio (CMRR) of Op-Amps is also very high, ranging from 60 to 140 dB.

Figure 5.2(a) illustrates the transfer characteristics of Op-Amps. Here, the voltage scale of the horizontal axis is much smaller than that of the vertical axis. The linear operating region is confined to a very narrow voltage range of the input signal. Beyond this region, the output signal becomes saturated, approaching the supply rail voltages (V_{S+} and V_{S-}). This linear input range can be estimated based on the

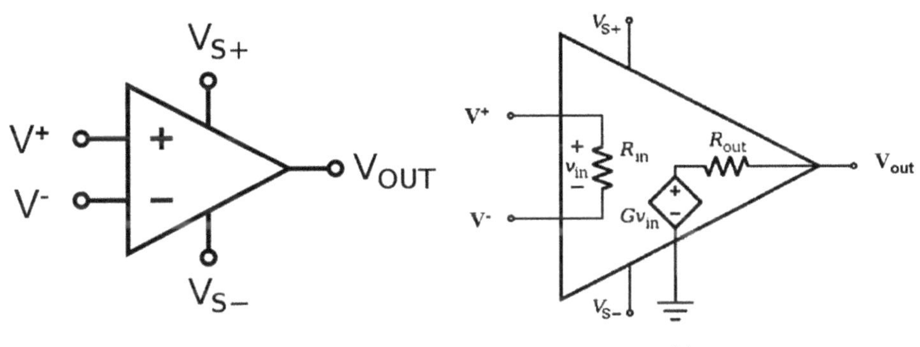

Figure 5.1. Operational amplifier: (a) symbol with pinout, (b) equivalent circuit. This [Op-Amp Internal] image has been obtained by the author from the Wikimedia website, where it is stated to have been released into the public domain. It is included within this book on that basis.

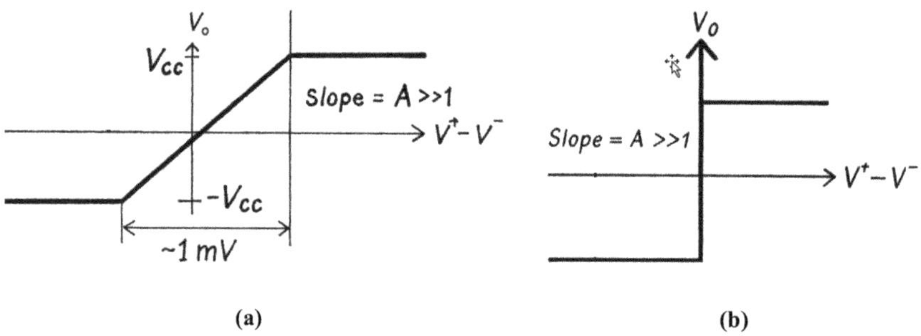

Figure 5.2. Transfer characteristics of an Op-Amp. Created with GPT-4.0, OpenAI.

voltage gain, which corresponds to the slope of the transfer curve. For example, if the rail voltages are ±10 V and the voltage gain is 10^5 V/V, then the input signal must be limited to less than ± 0.1 mV.

If the same voltage scale is used on both the horizontal and vertical axes, figure 5.2(a) is transformed into figure 5.2(b). From this diagram, an approximation can be made: $V_{in} = V^+ - V^- \approx 0$. This formula can be expressed in a simpler way:

$$V^+ \approx V^-. \tag{5.1}$$

It seems that the two input nodes are *virtually shorted*, a basic concept used in analyzing many amplifier circuits. Another approximation is that the input impedance goes to infinity, which implies that the input currents of Op-Amps are zero:

$$I^+ = I^- \approx 0. \tag{5.2}$$

The *virtual short* approximation, $V^+ \approx V^-$, is valid only with a negative feedback loop, which is present in most amplifier circuits. It works in this way: initially, there is a significant difference between the voltages at these two input nodes, but the output voltage does not jump to the rail voltage instantly. Instead, it takes a little time for the output voltage to change progressively. On the other hand, the response of the feedback path is much faster. When the condition of the *virtual short* is met, the output voltage stops changing, and the system remains in a stable state.

5.2 Simple Op-Amp circuits

5.2.1 Comparator circuit

The transfer characteristics shown in figure 5.2(b) demonstrate that a basic, open-loop Op-Amp can function as a *voltage comparator*. The corresponding circuit and its operation are illustrated in figure 5.3. In this configuration, two resistors form a voltage divider circuit that provides a reference voltage that is connected to the inverting input node. Alternatively, this reference can also be provided by an external voltage source.

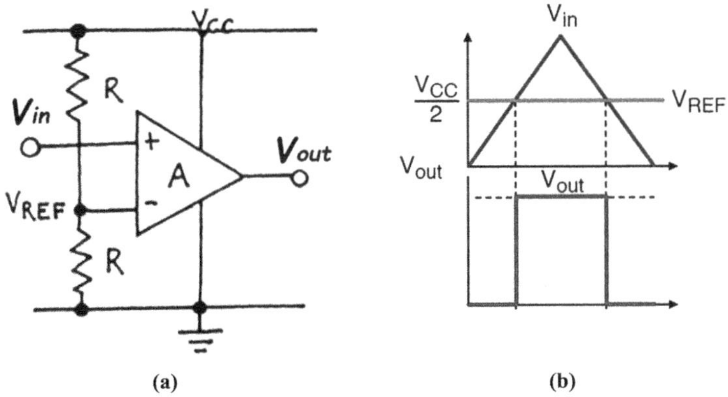

Figure 5.3. Voltage comparator circuit. Created with GPT-4.0, OpenAI.

In an Op-Amp comparator circuit, the output switches state based on a comparison between the input signal and a reference voltage. When the input voltage is lower than the reference voltage, the output swings toward the negative supply rail. Conversely, when the input voltage exceeds the reference, the output rises toward the positive supply rail. This sharp transition makes comparators useful for detecting threshold crossings with high sensitivity and speed.

One common application of comparators is in alarm and monitoring systems, where the reference voltage represents a critical threshold—for example, a maximum allowable temperature or pressure. When the monitored parameter crosses this threshold, the comparator output can trigger an alert, activate a relay, or initiate a control response. Comparators are also widely used in zero-crossing detectors, pulse-width modulation (PWM) circuits, and analog-to-digital conversion stages.

5.2.2 Buffer circuit

Another simple yet essential application of an Op-Amp is the voltage buffer, or unity-gain follower, as illustrated in figure 5.4(a). By applying the virtual short concept—where the voltage difference between the non-inverting and inverting inputs is nearly zero due to negative feedback—the output voltage closely tracks the input voltage, i.e. $V_{out} \approx V_{in}$.

The buffer serves several important functions. First, it facilitates unidirectional signal transfer, as shown in figure 5.4(b). In most situations, changes in subcircuit *B* do not affect subcircuit *A*, similar to the way a diode allows current flow primarily in one direction. Second, due to its extremely high input impedance, the buffer draws negligible current from subcircuit *A*, thereby minimizing loading effects and preserving signal integrity. Third, the buffer benefits from deep negative feedback, which significantly reduces its output impedance and enhances its bandwidth.

As a result of these characteristics, the buffer provides clean and stable signal transfer with minimal distortion.

One important application of Op-Amp buffers is in multistage amplifier circuits, where several amplifier stages are cascaded to achieve higher overall gain or specific frequency characteristics. In such configurations, buffer stages are used between

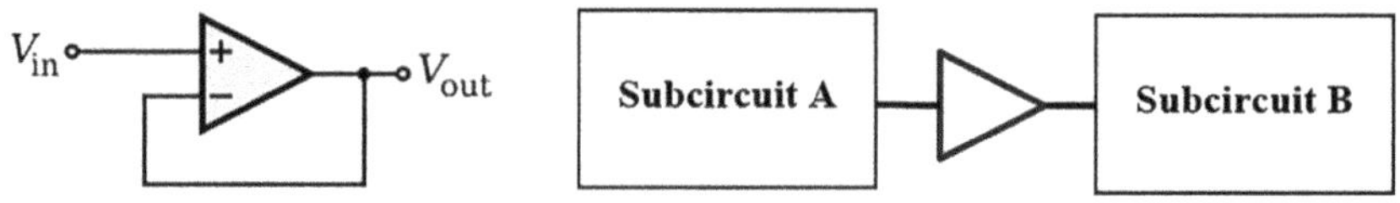

(a) (b)

Figure 5.4. Op-Amp buffer: (a) circuit, (b) application. This [Op-Amp Unity-Gain Buffer] image has been obtained by the author from the Wikimedia website, where it is stated to have been released into the public domain. It is included within this book on that basis.

amplifier blocks to maintain signal integrity by preventing loading effects. The high input impedance and low output impedance of the buffer ensure that the gain and frequency response of each stage remain unaffected by adjacent stages, thereby maximizing overall performance and stability.

Another key application of buffers is in oscillator circuits, where they play a crucial role in ensuring stable and reliable operation. Although the heart of an oscillator lies in its frequency-determining feedback network, the oscillator's output often needs to drive an external load. Without buffering, the load can interact with the feedback loop and disrupt the oscillation conditions, leading to frequency drift or instability. By isolating the oscillator core from the load, the buffer helps preserve the desired oscillation characteristics while enabling safe signal delivery to subsequent circuitry. In both applications, the buffer acts as a vital interface element, enabling modular circuit design and improving robustness in both analog signal processing and waveform generation systems.

5.2.3 Current–voltage converter circuit

Figure 5.5 illustrates a transimpedance amplifier (TIA) circuit implemented using an Op-Amp. This configuration is designed to convert an input current signal—typically from a photodiode or other current-generating sensor—into a corresponding voltage signal. Due to the Op-Amp's extremely high input impedance, it draws negligible current at its inverting input. As a result, virtually all of the input current flows through the feedback resistor R_f. Due to the virtual short principle ($V^+ \approx V^-$), the inverting input node is virtually grounded. Applying Ohm's law across the feedback resistor, the output voltage is given by: $V_{out} = -R_f I_{in}$. This negative sign indicates an inverting configuration, where the polarity of the output voltage is opposite to that of the input current.

Once the current signal is converted into a voltage, it can easily be amplified, filtered, or digitized by downstream signal processing circuits. In addition to signal conversion, TIAs also help improve signal-to-noise ratios and reduce loading effects, making them a critical component in high-sensitivity analog front-end designs.

TIAs are essential components in electronic systems where the primary signal is in the form of current rather than voltage. A prominent example is the photodiode, which generates a current proportional to the intensity of the incident light. Photodiodes are widely used across a range of applications, including optical communication systems, imaging devices, medical diagnostic equipment, and industrial automation systems involving optical sensors.

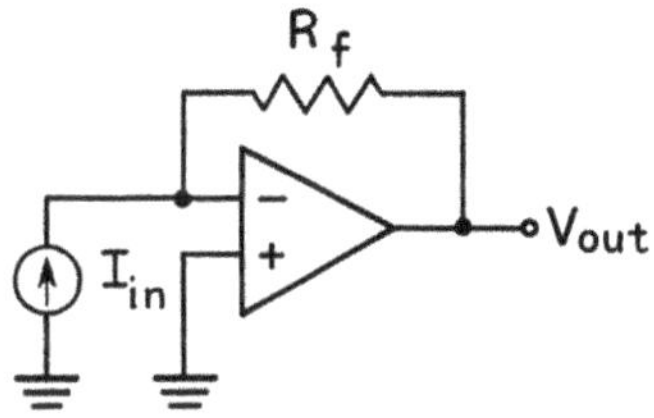

Figure 5.5. Transimpedance amplifier (TIA) circuit. Created with GPT-4.0, OpenAI.

Although the two inputs of an Op-Amp look similar, their functions are quite different in feedback circuits. The *inverting* input is typically used to establish negative feedback, which is essential for stable and predictable operation. For instance, in the circuit shown in figure 5.5, the feedback resistor must connect to the inverting input to ensure proper feedback control. If it were connected to the *non-inverting* input instead, the circuit would fail to operate correctly. Understanding the role of each input is critical in designing effective Op-Amp configurations.

5.3 Basic amplifier circuits

There are two basic configurations of Op-Amp amplifiers: the **inverting amplifier** and the **non-inverting amplifier**, each offering distinct characteristics. Together, these two amplifier types form the basis for many analog signal processing functions, including summing circuits, subtraction circuits, filters, differentiators, and integrators.

5.3.1 Inverting amplifier

Figure 5.6(a) illustrates the inverting amplifier circuit, where the input signal is applied to the inverting input through a resistor. According to the *virtual short* concept, the voltage at the inverting input node is held at zero volts—commonly referred to as a *virtual ground.* By applying Ohm's law to resistor R_1, the current flowing through the feedback path can be calculated. Then, applying Ohm's law to R_2, the relationship between the input and output voltages can be established, from which the voltage gain is derived:

$$A_V = \frac{V_{out}}{V_{in}} = -\frac{R_2}{R_1}. \tag{5.3}$$

For AC signals, the negative sign in the gain simply indicates a 180° phase shift, which is generally not a disadvantage in most applications. This formula indicates that the gain of this amplifier remains very stable across temperature variations. Transistors are known to be highly sensitive to temperature variations, whereas resistors are much more stable, with their resistance varying only slightly and typically in a linear manner with temperature: $R(T) = R_o[1 + \alpha(T - T_o)]$. Here, α is the *temperature coefficient of resistance*, which is less than 1% for most materials.

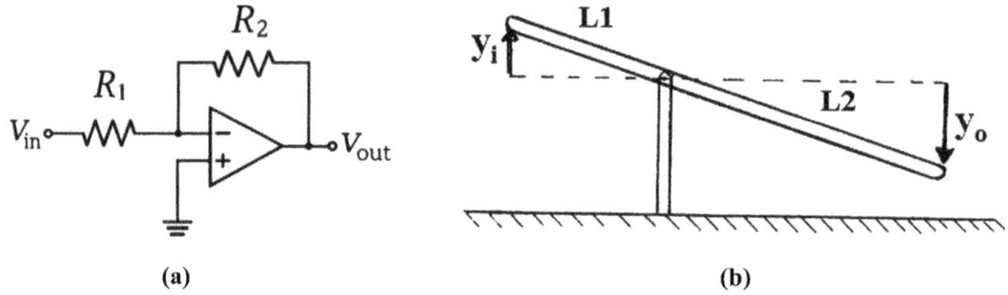

Figure 5.6. Inverting amplifier configuration: (a) circuit, (b) intuitive model. Created with GPT-4.0, OpenAI.

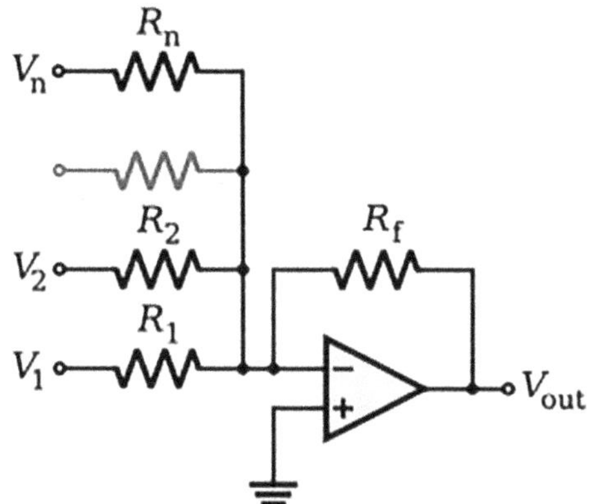

Figure 5.7. Summation amplifier circuit. This [Op-Amp IntegroSumming Amplifier] image has been obtained by the author from the Wikimedia website, where it is stated to have been released into the public domain. It is included within this book on that basis.

Moreover, the voltage gain expression in equation (5.3) depends on the ratio of two resistors. As a result, small temperature-induced changes in resistance tend to cancel out, provided the resistors have similar temperature coefficients.

Developing an intuitive understanding of amplifier behavior is highly beneficial. As illustrated in figure 5.6(b), this amplifier configuration resembles a seesaw, with the fulcrum representing the virtual ground in the circuit. The lengths L_1 and L_2 correspond proportionally to the resistor values R_1 and R_2, respectively. By applying the relationship between the two similar triangles in the diagram, the voltage gain formula can be derived.

The inverting amplifier configuration can be extended to accommodate multiple input signals, resulting in a **summing amplifier**, as shown in figure 5.7. According to Kirchhoff's current law (KCL), at the inverting input node, the current flowing through the feedback resistor R_f is equal to the sum of the currents from each input branch: $i_f = i_1 + i_2 + \ldots + i_n$. Using Ohm's law, the output voltage can then be expressed as a weighted sum of the input voltages:

$$V_{out} = -\left(\frac{R_f}{R_1}V_1 + \frac{R_f}{R_2}V_2 + \ldots + \frac{R_f}{R_n}V_n\right). \tag{5.4}$$

The primary application of this amplifier is in analog computation. For instance, computing a weighted sum is a fundamental operation in analog neural networks. If all the input resistors are equal to the feedback resistor, the output signal becomes the (negative) sum of all input signals. Additionally, this amplifier can also be used for *signal mixing*, including the combination of an AC signal with a DC offset.

5.3.2 Non-inverting amplifier

The primary drawback of the inverting amplifier is its low input impedance, which can be addressed by using the non-inverting configuration. As shown in figure 5.8(a), the input signal is applied directly to the non-inverting input of the Op-Amp. Based on the *virtual short* concept, the voltage at the inverting input node closely follows the input voltage. Since the Op-Amp draws negligible current, the two resistors in the feedback network form a voltage divider. The relationship between the input and output voltages is governed by this voltage divider, from which the voltage gain of the amplifier can be derived:

$$A_V = \frac{V_{out}}{V_{in}} = \frac{R_1 + R_2}{R_1} = 1 + \frac{R_2}{R_1}. \tag{5.5}$$

Since this formula is also expressed as a ratio of two resistors, the non-inverting amplifier exhibits high temperature stability as well.

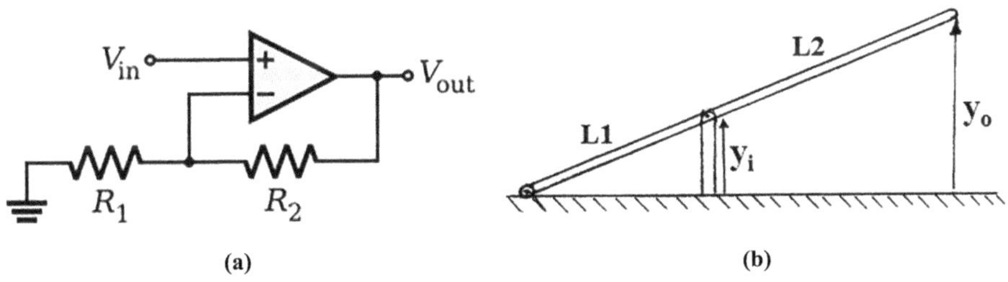

Figure 5.8. Non-inverting amplifier configuration: (a) circuit. This [Op-Amp Non-Inverting Amplifier] image has been obtained by the author from the Wikimedia website, where it is stated to have been released into the public domain. It is included within this book on that basis. (b) Intuitive model.

Figure 5.8(b) presents an intuitive model, analogous to a hydraulic crane. In this analogy, the height y_i represents the input signal, and the crane arm moves up and down in response. The lengths L_1 and L_2 are proportional to the resistor values R_1 and R_2, respectively. The diagram features two similar triangles, and the ratio of their sides reflects the same relationship described by equation (5.5).

Although the inverting and non-inverting configurations look quite different, they share an important characteristic: the feedback path is always connected to the inverting input of an Op-Amp. If the feedback path were mistakenly connected to the non-inverting input, the negative feedback would turn into positive feedback, resulting in completely different circuit behavior. Such a circuit with positive feedback is called a Schmitt trigger, which will be discussed in the last section of this chapter.

5.3.3 Subtraction amplifier

The inverting and non-inverting amplifier configurations can be combined to form a subtraction amplifier, which is shown in figure 5.9. The output signal can be determined using the superposition principle: $V_{out} = V_{o1} + V_{o2}$, where V_{o1} is the output due to the input signal V_1 when $V_2 = 0$, and V_{o2} is the output due to the input signal V_2 when $V_1 = 0$.

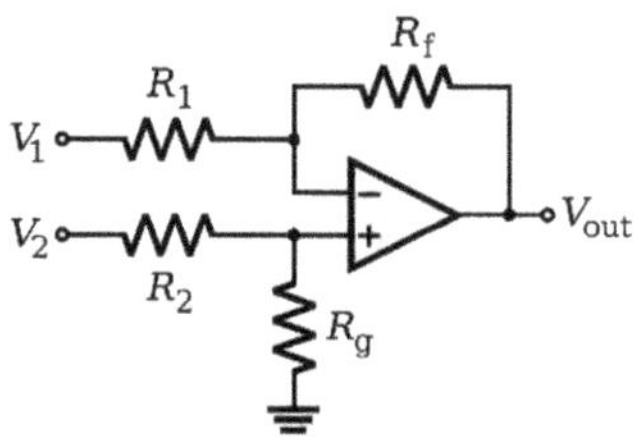

Figure 5.9. Subtraction amplifier circuit. This [Op-Amp Differential Amplifier] has been obtained by the author from the Wikimedia website, where it is stated to have been released into the public domain. It is included within this book on that basis.

When $V_2 = 0$, the non-inverting input node is grounded, and the circuit behaves as an inverting amplifier: $V_{o1} = -(R_f/R_1)V_1$. On the other hand, if $V_1 = 0$, it is not a simple non-inverting amplifier, since R_2 and R_g form a voltage divider. Taking this into account, the output is $V_{o2} = \frac{R_1 + R_f}{R_1}\frac{R_g}{R_2 + R_g}V_2$. If $R_1 = R_2$ and $R_f = R_g$, this expression is simplified to: $V_{o2} = (R_f/R_1)V_1$. Upon applying the superposition principle, the overall output voltage becomes:

$$V_{out} = \frac{R_f}{R_1}(V_2 - V_1). \tag{5.6}$$

As we mentioned at the beginning of this chapter, the core of an Op-Amp is a differential amplifier. However, it is often more convenient to use an Op-Amp to implement a differential amplifier. For example, this subtraction amplifier just needs four external resistors and provides reliable performance. One drawback, however, is its low input resistance, which can be resolved by adding Op-Amp buffer stages at the input nodes. In fact, there are integrated subtraction amplifiers that include the buffer stages, known as *instrumentation amplifiers*. As a result, for most applications requiring differential amplification, instrumentation amplifiers are the preferred solution.

5.4 First-order active filters

In section 1.4, we discussed first-order passive RC filter circuits, which can attenuate interference signals in the rejection band but are unable to amplify signals in the pass band. Active filters, which combine RC circuits with Op-Amps, overcome this limitation by allowing amplification within the pass band. Due to their precision, stability, and ease of design, active filters are widely used in audio processing, communication systems, and instrumentation.

5.4.1 Active low-pass filter

Figure 5.10(a) shows an active first-order low-pass filter (LPF) circuit with an inverting input configuration. The transfer function can be derived using the voltage gain formula for an inverting amplifier, where the capacitor C and resistor R_2 are combined into the impedance Z_2:

$$H(s) = \frac{\tilde{V}_{\text{out}}}{\tilde{V}_{\text{in}}} = -\frac{Z_2}{R_1} = -\frac{1}{R_1}\frac{R_2/sC}{R_2 + 1/sC} = -\frac{R_2}{R_1}\frac{\omega_c}{s + \omega_c}. \tag{5.7}$$

In this equation, the cutoff frequency is $\omega_c = 1/(R_2 C)$. The low-pass behavior of this filter can be intuitively understood by examining the two extreme cases: at very low frequencies, the capacitor behaves like an open circuit, effectively turning this circuit into an inverting amplifier with a gain of $H(s) = -R_2/R_1$. At high frequencies, the impedance of the capacitor becomes much less than R_2. As a result, R_2 is bypassed, and the transfer function drops progressively with increasing frequency.

Figure 5.10(b) depicts the Bode plot of this active filter with the parameters shown in the inset: $R_1 = 1$ kΩ, $R_2 = 10$ kΩ, and $C = 100$ nF. Using equation (5.7), the cutoff frequency and the gain in the pass band can be calculated as $f_c = 159$ Hz and $20 \log|A_v| = 20$ dB, respectively. These values closely match the simulation results. Additionally, the Bode plot clearly demonstrates the expected roll-off slope of -20 dB dec^{-1} in the high-frequency domain.

In addition to this inverting configuration, a non-inverting configuration can also be designed by combining a passive RC LPF and a non-inverting amplifier. However, compared to the inverting configuration, the non-inverting configuration is less compact, as it requires three resistors. For this reason, the inverting configuration shown in figure 5.10(a) is the preferred choice.

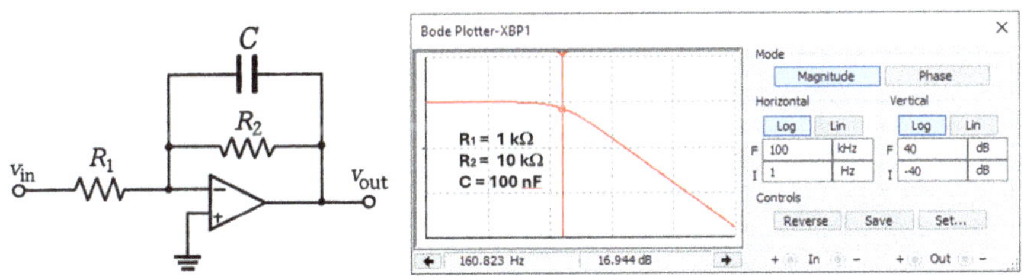

(a) (b)

Figure 5.10. First-order active low-pass filter (LPF): (a) circuit. Created with GPT-4.0, OpenAI. (b) Bode plot.

5.4.2 Active high-pass filter

Active high-pass filters (HPFs) can be constructed in a similar manner, as shown in figure 5.11(a). The transfer function can also be derived using the voltage gain formula for an inverting amplifier, where the capacitor C and resistor R_1 are combined into the impedance Z_1. The values of these two components determine the cutoff frequency: $\omega_c = 1/(R_1 C)$.

$$H(s) = -\frac{R_2}{Z_1} = -\frac{R_2}{R_1 + 1/sC} = -\frac{R_2}{R_1}\frac{s}{s + \omega_c} \tag{5.8}$$

Figure 5.11(b) depicts the Bode plot of this active filter with the parameters shown in the inset: $R_1 = 1$ kΩ, $R_2 = 10$ kΩ, C = 100 nF. Using equation (5.8), the cutoff frequency and the gain in the pass band are calculated as $f_c = 1.59$ kHz and $20\log|A_v| = 20$ dB, respectively. These values match the simulation results very well. In addition, the Bode plot clearly demonstrates the expected slope of 20 dB dec^{-1} in the low-frequency domain.

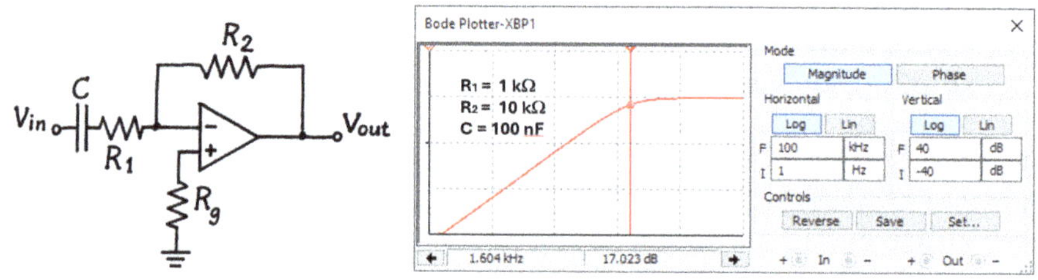

(a) (b)

Figure 5.11. First-order active high-pass filter (HPF): (a) circuit. Created with GPT-4.0, OpenAI. (b) Bode plot.

5.4.3 Active band-pass filter

The active LPF and HPF circuits can be combined to form a band-pass filter (BPF), as shown in figure 5.12(a). The two capacitors in this circuit have significantly different values, allowing for simplification in analysis. At high frequencies, the large capacitor C_1 behaves like a short circuit and thus can be ignored, making the circuit

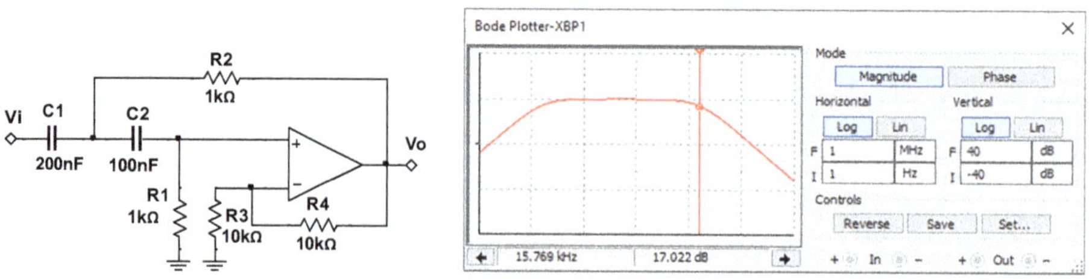

(a) (b)

Figure 5.12. Active band-pass filters (BPFs): (a) circuit, (b) Bode plot.

an LPF. At low frequencies, the small capacitor C_2 acts like an open circuit and thus can be disregarded, making the circuit an HPF. The transfer function can be derived in a similar way, i.e. by combining a capacitor and a resistor into an impedance:

$$H(s) = -\frac{Z_2}{Z_1} = -\frac{R_2}{R_1}\frac{s}{s+\omega_{c1}}\frac{\omega_{c2}}{s+\omega_{c2}}. \tag{5.9}$$

In this equation, $\omega_{c1} = 1/(R_1C_1)$, $\omega_{c2} = 1/(R_2C_2)$. Using the parameters provided in the circuit, the cutoff frequencies can be calculated: $f_{c1} = 15.9$ Hz, $f_{c2} = 15.9$ kHz. Figure 5.12(b) shows the Bode plot, which agrees with the calculated results very well.

Active filters offer several advantages over passive filters in the low- and medium-frequency domains. First, active filters use Op-Amps along with resistors and capacitors, eliminating the need for bulky and expensive inductors. Second, they provide gain, allowing signal amplification in addition to filtering, and offer better control over filter parameters such as cutoff frequency and quality factor. Third, active filters also maintain their performance, regardless of load impedance, and can implement higher-order filters with fewer components. These features make them ideal for compact, low-power, and precise analog signal processing in modern electronic systems.

5.5 Second-order active filters

First-order filters exhibit a gradual roll-off outside the pass band, which limits their effectiveness in suppressing interference signals that are close in frequency to the desired signal. To overcome this limitation, second-order or higher-order filters are typically employed. One straightforward approach is to cascade multiple first-order filter stages. However, this method presents certain challenges. Cascading active first-order filters requires multiple Op-Amps, increasing circuit complexity and power consumption. On the other hand, cascading passive first-order filters often results in a low Q-factor, which reduces the filter's selectivity. For example, directly cascading two first-order RC LPFs results in a second-order filter with a Q-factor of less than 0.5. Consequently, even signals below the cutoff frequency experience noticeable attenuation due to the gradual slope of the resulting transfer function.

The Sallen–Key topology is a popular filter configuration for second-order filters, where a positive feedback loop is used to boost the Q-factor. Introduced in 1955 by R P Sallen and E L Key at the MIT Lincoln Laboratory, Sallen–Key filters are a widely adopted family of active filters. Figure 5.13 illustrates a generic Sallen–Key filter circuit. For simplicity of analysis, the inverting input is shown as being connected directly to the output. In practical implementations, however, a feedback network consisting of two resistors is typically connected to the inverting input to provide gain within the pass band. However, the gain of this filter needs to be less than three; otherwise, it becomes unstable.

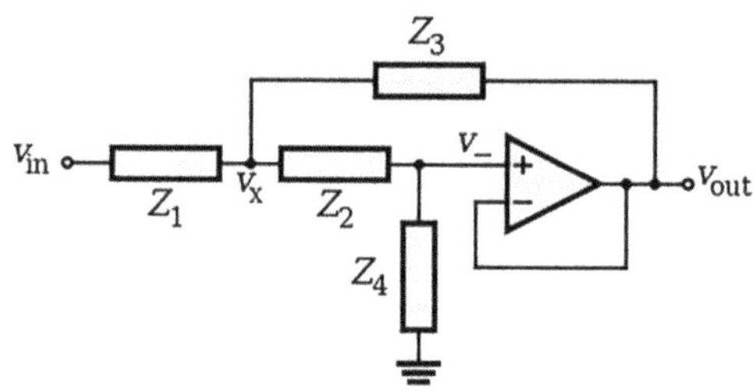

Figure 5.13. Generic Sallen–Key second-order active filter. This [Sallen-Key Generic Circuit] image has been obtained by the author from the Wikimedia website, where it is stated to have been released into the public domain. It is included within this book on that basis.

In the circuit shown in figure 5.13, $v_+ = v_- = v_o$, the only unknown node is $\tilde{V}_x$, which can be determined from the voltage divider circuit of Z_2 and Z_4:

$$\frac{\tilde{V}_-}{\tilde{V}_x} = \frac{Z_4}{Z_2 + Z_4} \rightarrow \tilde{V}_x = \frac{Z_2 + Z_4}{Z_4}\tilde{V}_o = \left(1 + \frac{Z_2}{Z_4}\right)\tilde{V}_o. \tag{5.10}$$

Applying KCL at the node V_x, the following equation can be set up:

$$\frac{\tilde{V}_i - \tilde{V}_x}{Z_1} = \frac{\tilde{V}_x - \tilde{V}_-}{Z_2} + \frac{\tilde{V}_x - \tilde{V}_o}{Z_3}. \tag{5.11}$$

The transfer function can be found by substituting equation (5.10) into equation (5.11):

$$H(s) = \frac{\tilde{V}_o}{\tilde{V}_i} = \frac{Z_3 Z_4}{Z_1 Z_2 + Z_1 Z_3 + Z_2 Z_3 + Z_3 Z_4}. \tag{5.12}$$

5.5.1 Sallen–Key low-pass filter

In the generic circuit shown in figure 5.13, the four external impedances can be grouped into two pairs: (Z_1, Z_3) and (Z_2, Z_4). Each pair forms a voltage divider configuration. By drawing on the principles of first-order passive RC filter circuits, it becomes straightforward to construct second-order active filters.

Figure 5.14(a) shows a circuit for a second-order LPF; its transfer function can be found by replacing the generic impedance with the specific values used in this circuit:

$$H(s) = \frac{\tilde{V}_o}{\tilde{V}_i} = \frac{1}{1 + C_2(R_1 + R_2)s + R_1 R_2 C_1 C_2 s^2} = \frac{\omega_c^2}{s^2 + 2\zeta\omega_c s + \omega_c^2}. \tag{5.13}$$

The critical frequency and damping factor can be determined as follows:

$$\omega_c = \frac{1}{\sqrt{R_1 R_2 C_1 C_2}}, \quad \zeta = \frac{1}{2}C_2(R_1 + R_2)\omega_c = \frac{1}{2}\sqrt{\frac{C_2}{C_1}}\left(\sqrt{\frac{R_1}{R_2}} + \sqrt{\frac{R_2}{R_1}}\right). \tag{5.14}$$

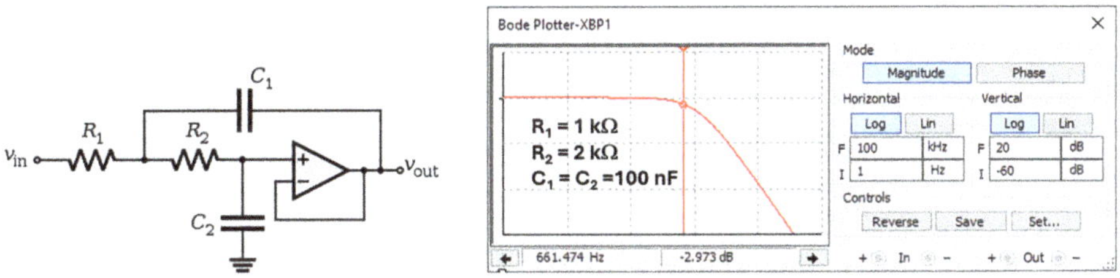

(a) (b)

Figure 5.14. Low-pass Sallen–Key filter: (a) circuit. This [Sallen-Key Lowpass Example] image has been obtained by the author from the Wikimedia website, where it is stated to have been released into the public domain. It is included within this book on that basis. (b) Bode plot.

At the critical frequency, the transfer function becomes $H(\omega_c) = -j/(2\varsigma) = -jQ$. Therefore, it is more convenient to determine the critical frequency from the phase plot, where the phase shift is $-90°$. On the other hand, at very low and very high frequencies, the expression for the transfer function can be simplified:

$$\begin{aligned} H(s) &\approx \frac{\omega_c^2}{\omega_c^2} = 1 \ (\text{for } \omega \ll \omega_c) \\ H(s) &\approx \frac{\omega_c^2}{s^2} = -\frac{\omega_c^2}{\omega^2} \ (\text{for } \omega \gg \omega_c). \end{aligned} \tag{5.15}$$

In the low-frequency domain ($\omega \ll \omega_c$), the capacitors are effectively open circuits, so the filter becomes a unity-gain buffer. Beyond the critical frequency, the transfer function curve in the Bode plot rolls off at the rate of -40 dB dec^{-1}.

Figure 5.14(b) presents the simulation results for the following parameters: $R_1 = 1$ kΩ, $R_2 = 2$ kΩ, and $C_1 = C_2 = 100$ nF. As expected, the Bode plot exhibits a roll-off rate of -40 dB dec^{-1}. However, the Q-factor is relatively low, $Q \approx 0.471$. Consequently, the cutoff frequency corresponding to a -3 dB drop is well below the critical frequency. Based on equation (5.14), the calculated critical frequency is $f_c = 1.125$ kHz. However, the cutoff frequency for a -3 dB drop is $f_{\text{cutoff}} = 661$ Hz, as indicated in the Bode magnitude plot in figure 5.14(b). Fortunately, the Q-factor can be boosted by increasing the gain of the filter, which is shown in figure 5.15.

In the circuit shown in figure 5.15(a), the voltage gain is set by resistors R_3 and R_4, resulting in a gain of $K = 2$ (V/V). Accordingly, the transfer function must be updated as shown in the following equation:

$$H(s) = \frac{\widetilde{V}_o}{\widetilde{V}_i} = K\frac{\omega_c^2}{s^2 + 2\zeta\omega_c s + \omega_c^2}. \tag{5.16}$$

In addition, the critical frequency and damping factor can be determined:

$$\omega_c = \frac{1}{\sqrt{R_1 R_2 C_1 C_2}}, \ \zeta = \frac{1}{2}\left[\sqrt{\frac{C_2}{C_1}}\left(\sqrt{\frac{R_1}{R_2}} + \sqrt{\frac{R_2}{R_1}}\right) - (K-1)\sqrt{\frac{C_1}{C_2}}\sqrt{\frac{R_1}{R_2}}\right]. \tag{5.17}$$

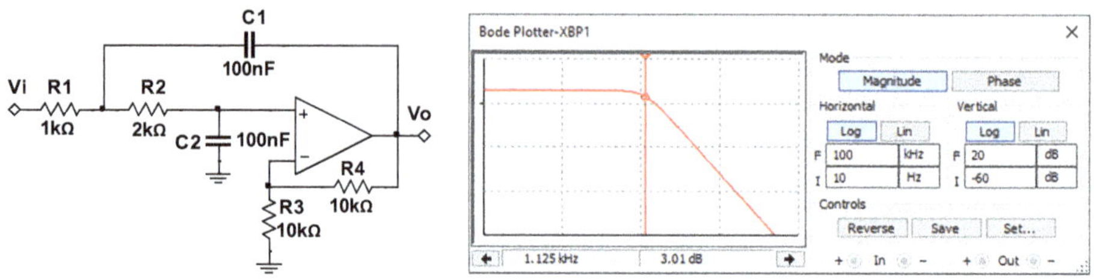

(a) (b)

Figure 5.15. Low-pass Sallen–Key filter with gain: (a) circuit, (b) Bode plot.

We observe that the expression for the critical frequency remains unchanged, but the damping factor and the Q-factor are different. Using the component values in the circuit from figure 5.15(a), the Q-factor is calculated as $Q = 1/\sqrt{2} \approx 0.707$, which corresponds to the condition for the maximum flat-frequency response (Butterworth filter). A clear improvement in filter performance is evident when directly comparing the two Bode plots in figure 5.14(b) and figure 5.15(b).

5.5.2 Sallen–Key high-pass filter

Similar to passive RC filters, an LPF can be converted into an HPF simply by interchanging the positions of the resistors and capacitors. Accordingly, figure 5.16 (a) illustrates an HPF configuration. The voltage gain in the pass band is determined by the resistors R_3 and R_4: $K = 1 + R_4/R_3$. Using the approach discussed earlier, the transfer function of this HPF can be derived as follows:

$$H(s) = \frac{V_O}{V_I} = K\frac{(R_1R_2C_1C_2)s^2}{(R_1R_2C_1C_2)s^2 + [R_2(C_1 + C_2) + (1 - K)R_1C_2] + 1}. \tag{5.18}$$

The denominator of this equation can be expressed in the standard format, as shown in equation (5.16), and then the two key parameters can be determined:

$$\omega_c = \frac{1}{\sqrt{R_1R_2C_1C_2}}, \quad \zeta = \frac{1}{2}\left[\sqrt{\frac{R_2}{R_1}}\left(\sqrt{\frac{C_1}{C_2}} + \sqrt{\frac{C_2}{C_1}}\right) - (K - 1)\sqrt{\frac{R_1}{R_2}}\sqrt{\frac{C_2}{C_1}}\right] \tag{5.19}$$

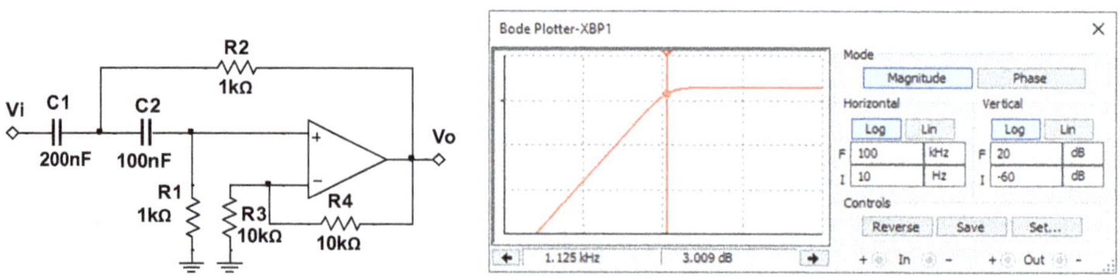

(a) (b)

Figure 5.16. High-pass Sallen–Key filter: (a) circuit, (b) Bode plot.

Figure 5.16(b) presents the Bode plot for an HPF with the following component values: $C_1 = 200$ nF, $C_2 = 100$ nF, $R_1 = R_2 = 1$ kΩ, and $R_3 = R_4 = 10$ kΩ. The calculated critical frequency is $f_c = 1.125$ kHz, and the Q-factor is 0.707, corresponding to the condition for the maximum flat-frequency response (Butterworth filter). The simulation results show good agreement with the theoretical prediction, thereby validating the derived equations.

> Due to the limited bandwidth of Op-Amps, the Bode plot eventually rolls off at higher frequencies. As a result, the pass band of the HPF does not extend beyond the Op-Amp's bandwidth. Consequently, with the frequency range extended, the overall frequency response resembles that of a BPF with a broad pass band.

5.5.3 Multiple feedback filters

In addition to Sallen–Key filter configurations, multiple feedback (MFB) filters also known as Deliyannis–Friend filters—are widely used in active filter design. In MFB filters, the non-inverting input of the Op-Amp is typically grounded, while the output is fed back to the inverting input through MFB paths. Figure 5.17 illustrates circuit implementations for both LPFs and HPFs using this topology. While both the Sallen–Key and MFB configurations provide excellent performance for these two types of filters, MFB designs generally require an additional resistor or capacitor in the feedback network. As a result, the Sallen–Key configuration is often preferred for these filter types due to its simpler structure and lower noise characteristics.

On the other hand, the MFB configuration is generally superior for designing BPFs. Figure 5.18(a) presents a typical MFB BPF circuit. In this design, the resistor R_2 is optional, but its inclusion provides additional flexibility in tuning the circuit's frequency response. Furthermore, the use of two identical capacitors is common practice, and it also leads to a simplified expression for the transfer function.

First, the horizontal capacitor and resistor R_3 are connected in series to the inverting input. Therefore, this part of the circuit is similar to an inverting amplifier, and the output voltage is related to the intermediate-node voltage V_1 through a

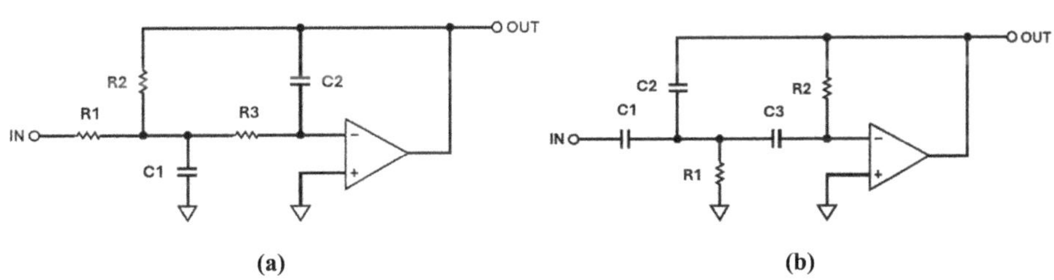

Figure 5.17. Multiple feedback (MFB) filters: (a) LPF, (b) HPF.

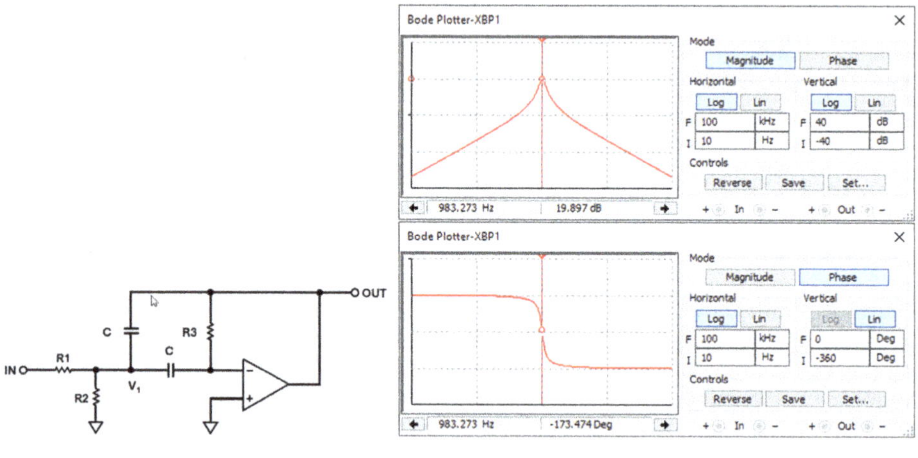

Figure 5.18. MFB BPF: (a) circuit, (b) Bode plot.

simple expression: $V_o = -sR_3CV_1$. Second, by applying KCL at the node with voltage V_1, the transfer function for this filter can be derived analytically:

$$H(s) = \frac{\widetilde{V}_o}{\widetilde{V}_i} = \frac{-s/(R_1C)}{s^2 + s(2/R_3C) + 1/[(R_1\|R_2)R_3C^2]}. \tag{5.20}$$

The standard transfer function of a BPF can be expressed as:

$$H(s) = K\frac{\omega_{BW}s}{s^2 + \omega_{BW}s + \omega_o^2}. \tag{5.21}$$

In this equation, K stands for the voltage gain at the central frequency, while ω_o and ω_{BW} denote the central frequency and bandwidth, respectively. By comparing these two equations, the three key parameters can be determined: $\omega_o = 1/[C\sqrt{(R_1\|R_2)R_3}]$, $\omega_{BW} = 2/R_3C$, and $K = -R_3/(2R_1)$. Additionally, the quality factor can be calculated; it is the ratio of the central frequency to the bandwidth: $Q = \sqrt{R_3/(R_1\|R_2)}/2$. If a high quality factor is desired, R_3 should be much higher than R_1 and R_2: $R_3 \gg R_1\|R_2$. In addition, it is interesting to note that the quality factor is independent of the capacitor.

Figure 5.18(b) shows the simulation results for the following component values: $R_1 = 8$ kΩ, $R_2 = 2$ kΩ, $R_3 = 160$ kΩ, and $C_1 = C_2 = 10$ nF. The central frequency and the gain at the peak are observed to be $f_o = 983$ Hz and $K = -9.88$ V/V, respectively. The negative sign is indicated in the phase plot, where a phase shift of $-180°$ is shown at the central frequency. The bandwidth can also be determined from the Bode plot: $f_{BW} = f_H - f_L = 244$ Hz, where the upper and lower cutoff frequencies are found to be $f_H = 1125$ Hz and $f_L = 881$ Hz, respectively. These simulation results agree reasonably well with the theoretical calculations, with minor discrepancies attributed to the limited resolution of the simulation tool.

5.6 Introduction to negative feedback

In a system consisting of two functional blocks, there are three primary ways to interconnect them: series, parallel, and feedback configurations. Figure 5.19 illustrates a block diagram of a basic feedback system. From this diagram, the relationships among the four key parameters can be derived:

$$\begin{aligned} X_f(s) &= G_2(s)X_o(s) \\ X_i(s) &= X_s(s) \pm X_f(s) \\ X_o(s) &= G_l(s)X_i(s). \end{aligned} \tag{5.22}$$

There are two options in the second equation: $X_i(s) = X_s(s) + X_f(s)$ for positive feedback and $X_i(s) = X_s(s) - X_f(s)$ for negative feedback, which is shown in figure 5.19.

Consider the system enclosed within a black box, such that only the source and output signals are observable from the outside. In this context, the transfer function is defined as the ratio of the output to the source, which can be derived using equation (5.22):

$$H(s) = \frac{X_o(s)}{X_s(s)} = \frac{G_1(s)}{1 \mp G_1(s)G_2(s)}. \tag{5.23}$$

For a system with negative feedback, the denominator takes the form $1 + G_1(s)G_2(s)$. Assuming $G_1(s)G_2(s)$ is positive, the overall gain of the system is reduced compared to the open-loop configuration without feedback. However, the inclusion of a negative feedback loop in electronic circuits offers numerous significant advantages. It enhances system stability, improves linearity, reduces sensitivity to component variations, and lowers noise. In amplifier circuits specifically, negative feedback can extend the bandwidth, reduce total harmonic distortion (THD), and improve both input and output impedance characteristics. For voltage amplifiers, the input impedance is often increased, minimizing the loading of the signal source, while the output impedance is reduced, improving the circuit's ability to drive loads. Additionally, negative feedback enhances power supply rejection and improves overall signal fidelity, making it a fundamental design technique in high-performance analog systems.

The non-inverting amplifier circuit is a simple example of a negative feedback system, which is shown in figure 5.20. In this circuit, the two resistors form the feedback network, and the feedback gain can be determined using the voltage divider formula:

$$\beta = \frac{X_f}{X_o} = \frac{V_-}{V_o} = \frac{R_1}{R_1 + R_2}. \tag{5.24}$$

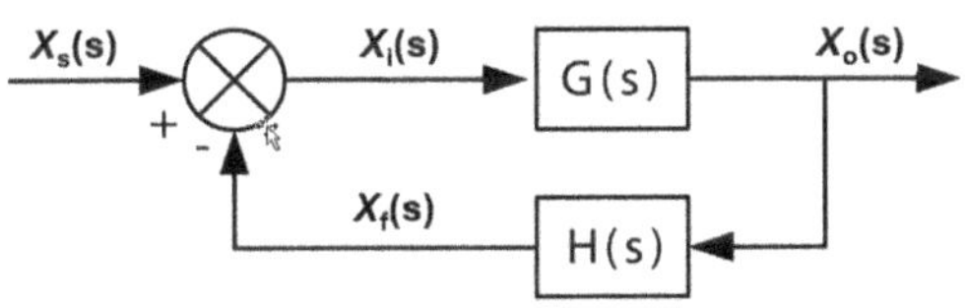

Figure 5.19. System diagram with negative feedback.

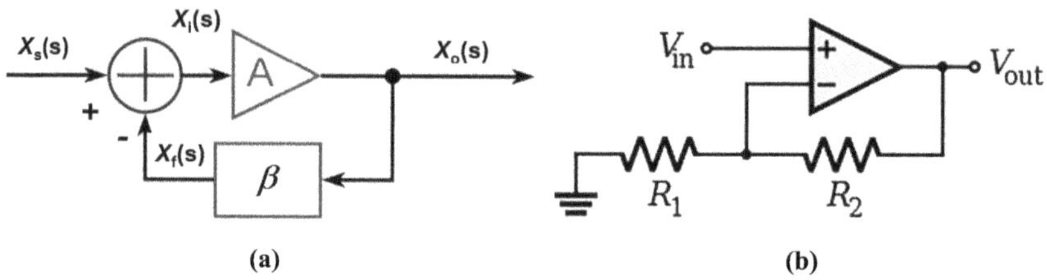

Figure 5.20. Non-inverting amplifier: (a) block diagram, (b) circuit.

From the perspective of the Op-Amp, the effective input signal is the differential voltage between the non-inverting and inverting input terminals, denoted by $V_i = V_+ - V_-$. This differential input is amplified by the Op-Amp's open-loop gain: $V_o = AV_i$. Since the open-loop gain of Op-Amps is very high, $A\beta \gg 1$ is a good approximation. It turns out that the derived closed-loop gain, as shown in equation (5.25), does not depend on the open-loop gain of the Op-Amp. This result confirms the robustness and predictability of feedback amplifier design, and it aligns with the gain expression derived through direct circuit analysis:

$$A_f = \frac{A}{1 + A\beta} \approx \frac{1}{\beta} = \frac{R_1 + R_2}{R_1}. \tag{5.25}$$

In the previous chapter, we discussed stability and compensation techniques for Op-Amps. As a trade-off for ensuring stability, the dominant pole is intentionally placed at a low frequency. For instance, in the widely used LM741 Op-Amp, the dominant pole is located at around 10 Hz, making it unsuitable for most applications in open-loop configurations. However, when negative feedback is applied, the frequency response of the system improves dramatically.

At the cost of reduced gain, negative feedback can extend the bandwidth by effectively pushing the closed-loop pole to a higher frequency. Assuming the open-loop response of the Op-Amp can be approximated by a single-pole transfer function, the frequency response of the resulting closed-loop feedback system can be analytically derived:

$$A(s) = A_o \frac{\omega_c}{s + \omega_c} \Rightarrow A_f(s) = \frac{A(s)}{1 + A(s)\beta} = \frac{A_o}{1 + A_o\beta} \frac{\omega_c(1 + A_o\beta)}{s + \omega_c(1 + A_o\beta)}. \tag{5.26}$$

In this context, the parameter known as the *amount of feedback* $(1 + A_o\beta)$ plays a critical role in shaping the performance of a feedback amplifier. First, the application of negative feedback reduces the DC gain of the system by this factor: $A_{fo} = A_o/(1 + A_o\beta)$. Second, and more importantly, it extends the bandwidth of the amplifier by the same factor: $\omega_{fc} = \omega_c(1 + A_o\beta)$.

Figure 5.21 illustrates the inverse relationship between gain and bandwidth in Op-Amp circuits at different levels of negative feedback. The Op-Amp has an open-loop DC gain of 100 dB with a dominant pole at 10 Hz, resulting in a narrow bandwidth. When feedback is applied, the gain drops but the bandwidth widens. For example, with the amount of feedback set to $1 + A_o\beta = 10$, the gain is reduced to 80 dB, and

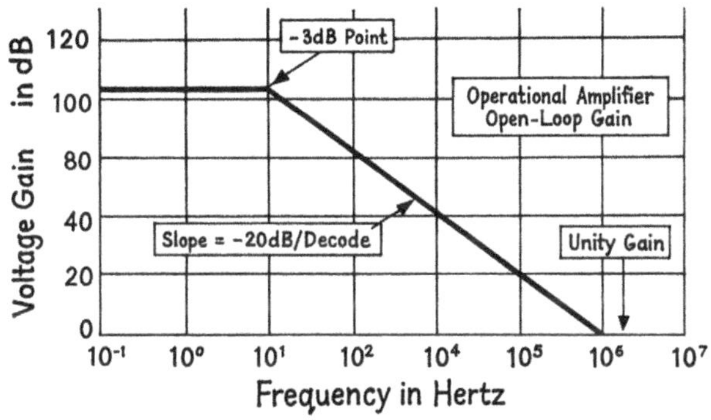

Figure 5.21. Frequency response of an Op-Amp with feedback. Created with GPT-4.0, OpenAI.

the bandwidth increases to approximately 100 Hz. If the amount of feedback is increased to 10^4, the gain drops to 20 dB, and the bandwidth extends to approximately 100 kHz. In the extreme case of a unity-gain buffer, the closed-loop gain is 0 dB, and the bandwidth reaches its maximum at about 1 MHz.

> The frequency response behavior shown in figure 5.21 highlights an important property: the product of gain and bandwidth remains approximately constant. This value, known as the *gain–bandwidth product* (GBP or GBW), is an intrinsic figure of merit for an Op-Amp. Furthermore, when the gain is reduced to unity (0 dB) under deep feedback, the resulting bandwidth is equal to this *gain–bandwidth product*. Hence, the *unity-gain frequency* and *gain–bandwidth product* refer to the same parameter, representing the maximum frequency at which the Op-Amp can operate with a closed-loop gain of one.

Figure 5.22 shows the simulation results for a non-inverting amplifier configuration. The Op-Amp used has the same characteristics as the one shown in figure 5.21: $A_o = 10^5$ V/V (100 dB) and $f_c = 10$ Hz. The feedback factor is determined by the values of the two resistors: $\beta = R_1/(R_1 + R_2) = 10^{-3}$; and then the amount of feedback is calculated: $1 + A_o\beta = 101$. Using equation (5.26) the closed-loop gain and bandwidth of the feedback amplifier can be calculated: $A_V \approx 990$ V/V (59.9 dB), and $f_{fc} = 1.01$ kHz.

The Bode plot shown in figure 5.22 demonstrates excellent agreement with the theoretical predictions, confirming the relationship between the gain, bandwidth, and the amount of feedback in a non-inverting amplifier. This result also reinforces the concept that negative feedback extends bandwidth at the expense of gain while preserving the *gain–bandwidth product*.

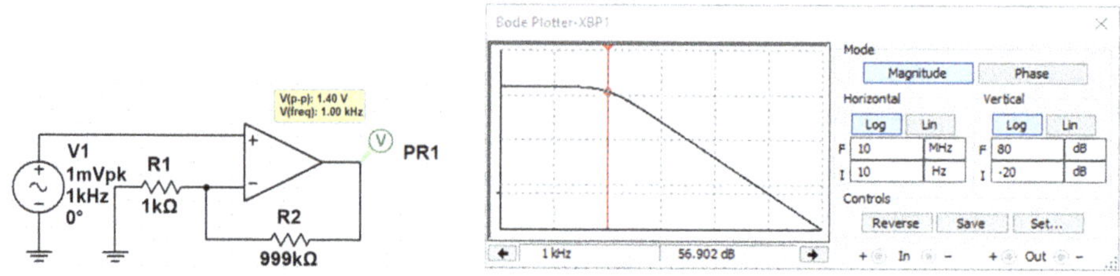

(a) (b)

Figure 5.22. Non-inverting amplifier. DC gain ≈ 59.9 dB, BW ≈ 1 kHz.

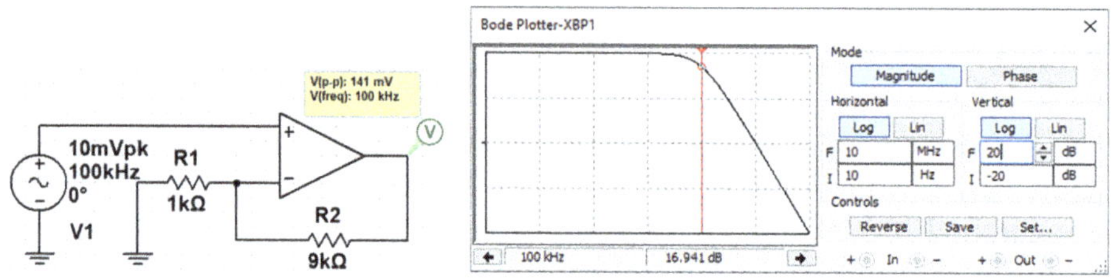

(a) (b)

Figure 5.23. Non-inverting amplifier. DC gain ≈ 20 dB, bandwidth ≈ 100 kHz.

Figure 5.23 shows the simulation results for a similar circuit, but with a significantly increased feedback factor, $\beta = 0.1$. This results in an *amount of feedback* equal to $1 + A_0\beta = 10001 \approx 10^4$. This deep negative feedback reduces the closed-loop gain to 10 V/V (20 dB), but the bandwidth is extended to 100 kHz. The Bode plot presented in figure 5.23 confirms this behavior, showing reduced gain and extended bandwidth consistent with theoretical predictions.

> In general, negative feedback makes an amplifier's frequency response more predictable, flatter, and wider, allowing it to accurately amplify a broader range of signals with less distortion. This is crucial for high-fidelity audio, precise instrumentation, and various other electronic applications.

5.7 Improving insensitivity using negative feedback

Op-Amps are composed of transistors, which are inherently very sensitive to temperature variations. As a result, their open-loop gain can fluctuate significantly with temperature changes. However, when a negative feedback loop is introduced, the overall closed-loop gain becomes largely determined by the ratio of the external resistors—components that are typically more thermally stable. This enhances the insensitivity of the circuit.

The improvement in insensitivity can be analyzed by taking the derivative of the closed-loop gain expression with respect to the open-loop gain. This yields the following relationship:

$$A_\mathrm{f} = \frac{A}{1 + A\beta} \Rightarrow \frac{dA_\mathrm{f}}{A_\mathrm{f}} = \frac{1}{1 + A\beta}\frac{dA}{A}. \tag{5.27}$$

For example, suppose the open-loop gain of the Op-Amp varies significantly with temperature, $dA/A = 50\%$. When $1 + A\beta \geqslant 100$, the percentage change of the closed-loop gain is very limited: $dA_\mathrm{f}/A_\mathrm{f} \leqslant 0.5\%$. This demonstrates how negative feedback can significantly reduce the sensitivity of an amplifier's performance to variations in both internal and environmental factors.

In many precision instruments, a stable supply voltage is essential to ensure proper functionality and to maintain the specified performance parameters. Fluctuations in the supply voltage can lead to measurement errors, drift, and degraded signal integrity. To address this issue, voltage regulators are widely employed. These circuits are designed to maintain a constant output voltage despite variations in input voltage, load current, or ambient temperature—provided these variations remain within the regulator's specified operating range.

Figure 5.24(a) illustrates the block diagram of a voltage regulator with several key functional elements, and an example of a voltage regulator circuit is shown figure 5.24(b). First, a stable *reference voltage* is essential; in the example circuit, a Zener diode is used for this purpose, though it can be replaced with a more precise reference subcircuit, such as a bandgap reference, for improved stability.

Second, the output voltage is sampled, which provides the *feedback* signal. In the example circuit, the output signal is taken directly as the feedback signal. However, a voltage divider parallel to the load is often used to offer design flexibility in setting the desired output level. For example, given a reference voltage of 5 V, the output voltage can be 10 V if the feedback signal is taken between two identical resistors in series. To avoid bypassing too much current, the values of these two resistors should be quite large.

Third, an Op-Amp serves as an *error detection* device by comparing the scaled output voltage with the reference voltage and generating an error signal. Finally, a

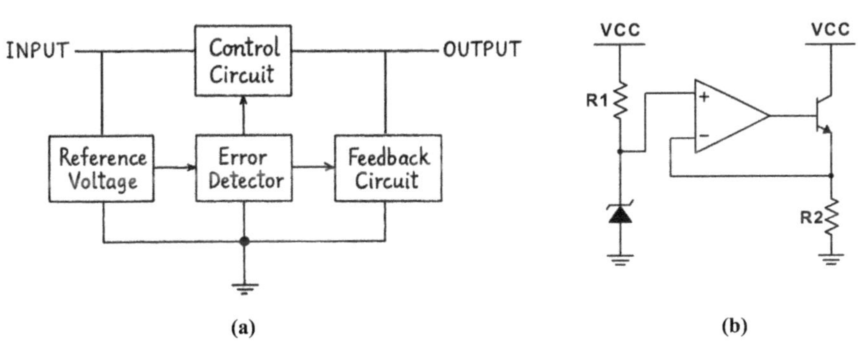

Figure 5.24. Voltage regulator: (a) block diagram, (b) example circuit. Created with GPT-5.0, OpenAI.

power transistor acts as the *control circuit* device, regulating the current delivered to the load based on the error signal. Together, these components form a closed-loop system that maintains a stable and regulated output voltage, even under varying input voltage and load demands.

As introduced in chapter 3, Multisim provides a powerful sensitivity analysis feature, and the results are presented in figure 5.25. Assuming that the Op-Amp operates ideally and reliably, this analysis focuses on the remaining factors that can influence the circuit's performance. For a **BJT** transistor, two primary sources of variation are considered: the junction area and temperature. The numerical values shown in the sensitivity table in figure 5.25(b) represent partial derivatives, which quantify how the output voltage changes with respect to small variations in each component parameter. Mathematically, these values correspond to the derivatives in the following sensitivity equation, which is used to assess how sensitive the output is to fluctuations in device characteristics or environmental conditions:

$$\mathrm{d}V = \frac{\partial V}{\partial A}\mathrm{d}A + \frac{\partial V}{\partial T}\mathrm{d}T + \sum_k \frac{\partial V}{\partial R_\mathrm{k}}\mathrm{d}R_\mathrm{k} + \frac{\partial V}{\partial V_\mathrm{CC}}\mathrm{d}V_\mathrm{CC} + \frac{\partial V}{\partial V_\mathrm{EE}}\mathrm{d}V_\mathrm{EE}. \tag{5.28}$$

Among the data obtained from the sensitivity analysis shown in figure 5.25(b), the resistor R_2 (rr2) and the temperature of the transistor (qq1_temp) exhibit the least sensitivity, with a value in the range of 10^{-9}, indicating a negligible influence on the output voltage. Following this, the junction area of the **BJT** (qq1_area), which typically exhibits minimal variation during fabrication processes, contributes to sensitivity in the order of 10^{-7}. The sensitivity to the value of R_1 (rr1) is slightly higher, in the range of 10^{-6}, and the negative sign indicates an inverse correlation—i.e. an increase in $R1$ results in a slight decrease in output voltage. The most influential factor is the supply voltage, V_{CC}, with a sensitivity of approximately 5.17×10^{-3}.

To validate and better understand this result, an AC voltage source is superimposed on the $\mathrm{V_{CC}}$ rail, as shown in figure 5.26. When a 1 V AC signal is applied to $\mathrm{V_{CC}}$, the resulting output voltage ripple has an amplitude of approximately 5.25 mV, corresponding to a sensitivity of 5.25×10^{-3}. This empirical result closely aligns with

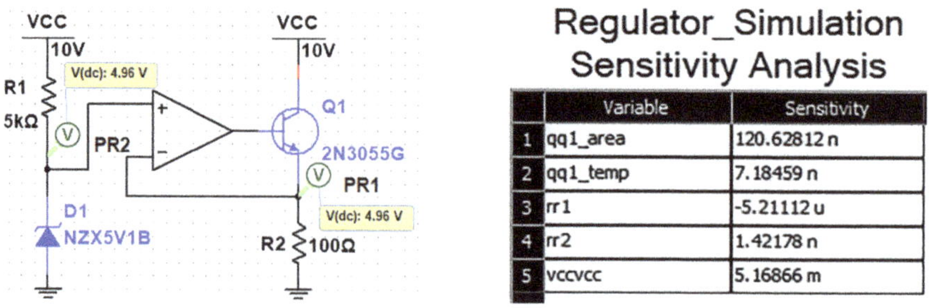

	Variable	Sensitivity
1	qq1_area	120.62812 n
2	qq1_temp	7.18459 n
3	rr1	-5.21112 u
4	rr2	1.42178 n
5	vccvcc	5.16866 m

(a) (b)

Figure 5.25. Voltage regulator: (a) circuit, (b) sensitivity analysis.

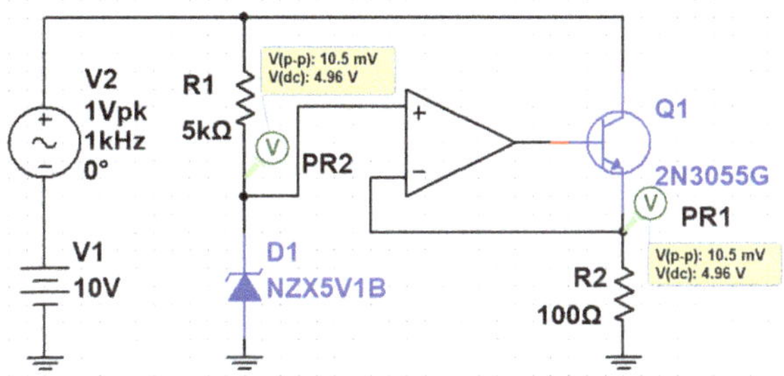

Figure 5.26. Direct sensitivity analysis of the voltage regulator circuit.

the sensitivity value (5.17×10^{-3}) listed in figure 5.25(b), providing confirmation of the analysis.

In figure 5.26, an additional probe is placed at the top of the Zener diode to measure both the DC voltage and the AC signal. The probe readings reveal that the voltage at this node is identical to that of the output probe, confirming that both the DC output voltage and the AC (sensitivity) response are governed by the Zener diode. This indicates that the Zener diode serves as the primary reference that not only establishes the nominal voltage but also determines how the output responds to variations in environmental factors. As a result, the stability of the output is directly tied to the characteristics of the Zener diode. To achieve improved performance—particularly in applications requiring higher precision or lower temperature sensitivity—the Zener diode can be replaced with a more stable voltage reference, such as a bandgap reference circuit, which offers better regulation with respect to temperature and supply voltage variations.

5.8 Improving linearity using negative feedback

Linearity in electronic circuits refers to the ability of a system—typically an amplifier—to produce an output signal that is directly proportional to its input. In real-world circuits, nonlinearity arises due to the inherently nonlinear characteristics of active components such as transistors or Op-Amps. Negative feedback is a fundamental technique used to significantly improve linearity in such systems.

When a purely sinusoidal input signal at a frequency of f_o passes through a nonlinear circuit, the output is not usually a perfect sine wave. Instead, it contains components at the fundamental frequency with an amplitude of V_1 and at higher harmonic frequencies ($2f_o$, $3f_o$, $4f_o$...) with amplitudes of V_2, V_3, V_4, etc. As introduced in chapter 3, THD is a metric widely used to quantify nonlinearity in electronic systems, especially amplifiers, signal processing circuits, and communication devices. It provides a measure of how much a system distorts a pure input signal by generating additional harmonic components due to its nonlinear characteristics. THD is defined in this way:

$$THD = \frac{\sqrt{V_2^2 + V_3^2 + V_4^2 + \dots}}{V_1}. \tag{5.29}$$

THD is often presented as a percentage, since it is usually quite small. The requirement for linearity varies widely for different applications. In the area of power electronics, THD is required to be less than 5%. For general-purpose amplifiers, THD is required to be less than 1%. In addition, for Hi-Fi audio amplifiers, THD is required to be less than 0.1%.

A push-pull power amplifier (PA), which belongs to the family of Class B PAs, is a commonly used configuration in analog electronics, particularly for efficient amplification of audio and RF signals. It achieves higher output power by employing two transistors that alternately amplify the positive and negative halves of an input sine wave. However, due to the threshold voltage required for the transistors to conduct, crossover distortion can be significant—especially in a basic amplifier configuration like the one shown in figure 5.27(a). Multisim provides a convenient virtual tool for measuring THD, which reaches 23.1%, as shown in the circuit diagram. While low levels of distortion may be difficult to detect visually, figure 5.27 (b) clearly illustrates the pronounced distortion near the zero-crossing points of the waveform.

The distortion in this circuit can be effectively eliminated by applying negative feedback, as shown in figure 5.28. In general, power amplifiers are designed to amplify current rather than voltage, meaning the input and output signals often have similar voltage amplitudes. As a result, the output voltage of the power amplifier can be directly fed back to the inverting input of the Op-Amp. This feedback loop compels the system to correct deviations from the input signal, producing an output waveform that exhibits minimal distortion.

The improvement in linearity achieved through negative feedback seems almost magical, yet it often leaves learners puzzled about how it works. To gain deeper insight, the output signal of the Op-Amp—prior to the power amplifier stage—is examined in the simulation shown in figure 5.29. The results reveal that this signal is

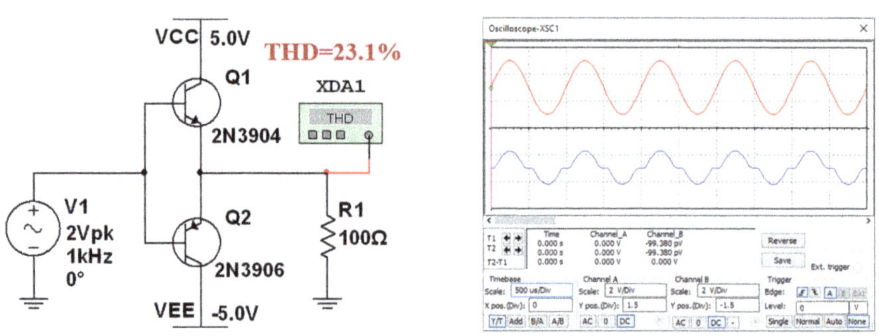

(a) (b)

Figure 5.27. Push-pull power amplifier: (a) circuit, (b) simulation results.

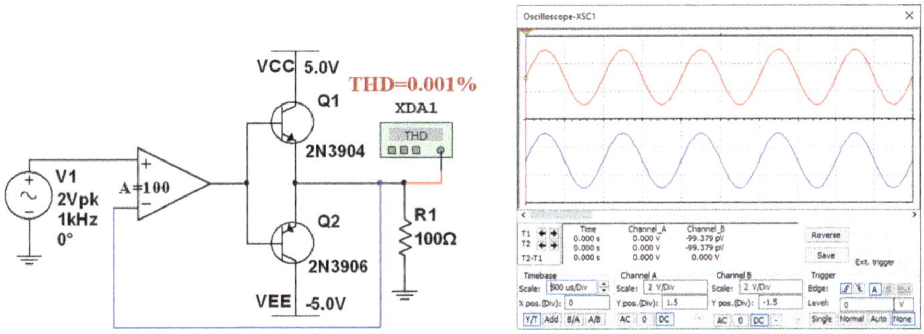

(a) (b)

Figure 5.28. Push-pull power amplifier with feedback: (a) circuit, (b) simulation result.

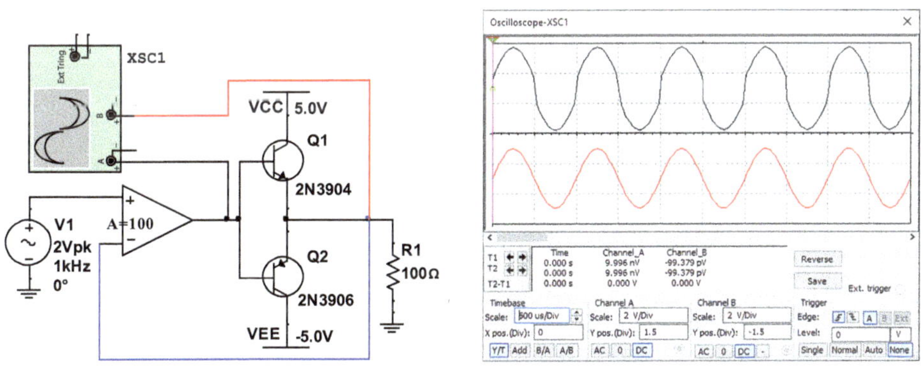

(a) (b)

Figure 5.29. Investigation of feedback: (a) circuit, (b) simulation results.

intentionally *pre-distorted* by the Op-Amp. This pre-distortion effectively compensates for the nonlinear behavior—particularly the crossover distortion—of the push-pull power stage. As a result, once the signal passes through the nonlinear output stage, the final output becomes a highly linear reproduction of the original input. This illustrates a key principle of negative feedback: it shapes the internal signals of a system in such a way that the overall output matches the desired response, even when nonideal components are involved.

To further illustrate the concept of crossover distortion, we examine the super-diode circuit. Figure 5.30 presents a basic half-wave rectifier along with its simulation results. Due to the characteristics of a *pn* junction diode, its current remains negligibly small when the forward bias voltage is below approximately 0.6 V. As a result, the output voltage remains near zero whenever the input signal is below this threshold. Conceptually, if a horizontal line is drawn at $V = 0.6$ V in

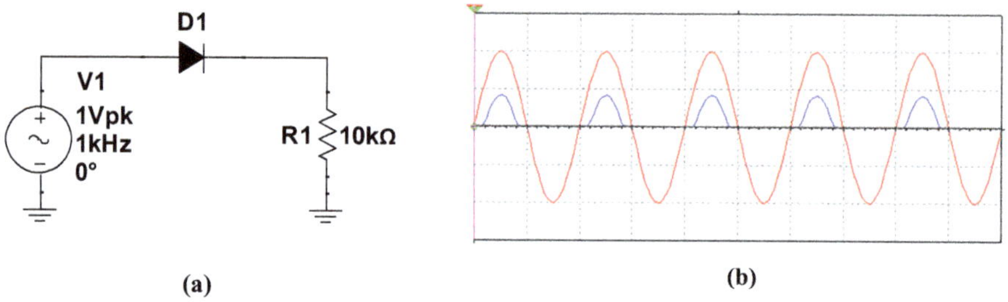

Figure 5.30. Half-wave rectification using a diode: (a) circuit, (b) simulation results.

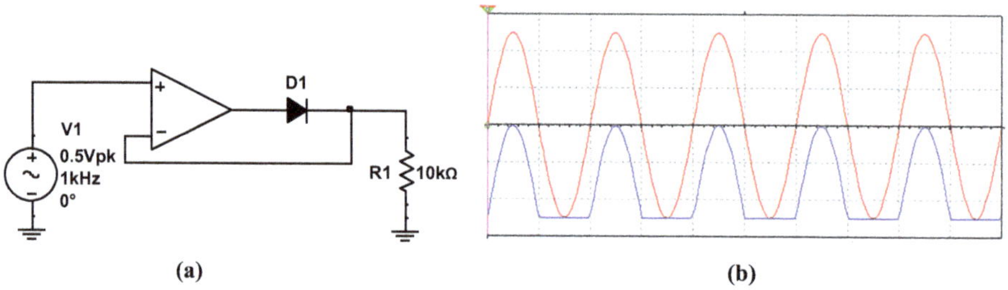

Figure 5.31. Half-wave rectification using a superdiode: (a) circuit, (b) simulation results.

figure 5.30(b), only the portion of the input waveform above this line appears at the output. This illustrates the nonlinear behavior introduced by the diode's turn-on threshold. In addition, if the input signal is quite weak, with an amplitude of less than 0.6 V, the output signal is hard to obtain.

A feedback loop is now introduced, as shown in figure 5.31(a), which forces the output signal to closely track the input signal during the conducting phase. In addition, the amplitude of the input signal is reduced to 0.5 V. To clearly visualize the input and output waveforms in figure 5.31(b) and avoid overlap, the output signal is shifted downward by 0.5 V for display. This precision rectifier is capable of accurately rectifying very weak AC signals, making it highly valuable in sensitive analog applications. Such a combination of an ordinary diode with an Op-Amp forms a 'superdiode,' which has ideal rectifying behavior.

> There are many applications of the 'superdiode,' and the precision rectifier discussed in this section is just one of them. In addition, they are also widely used in the fields of active filters, audio electronics, analog signal processing, measurement, instrumentation, etc.

5.9 Positive feedback in Op-Amp circuits

Systems with negative feedback tend to be stable, as illustrated in figure 5.32(a). When a disturbance occurs, the ball is displaced from its equilibrium position, but a restoring force acts to bring it back. In contrast, systems with positive feedback are inherently unstable, as shown in figure 5.32(b), because any perturbation results in a force that drives the ball further away from the equilibrium position.

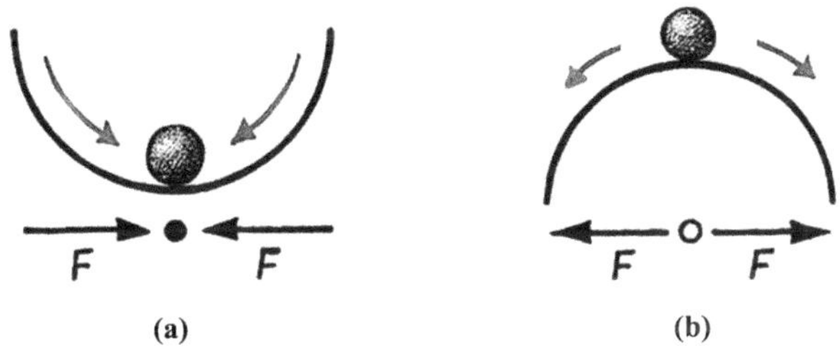

Figure 5.32. Equilibria: (a) stable, (b) unstable. Created with GPT-4.0, OpenAI.

If the feedback network, consisting of two resistors in series, is connected to the non-inverting input of the Op-Amp, it creates a positive feedback loop, causing the output to approach one of the supply rails—either the positive or negative voltage limit. Figure 5.33 illustrates two such circuit configurations. Their corresponding negative feedback counterparts are the well-known inverting and non-inverting amplifier circuits.

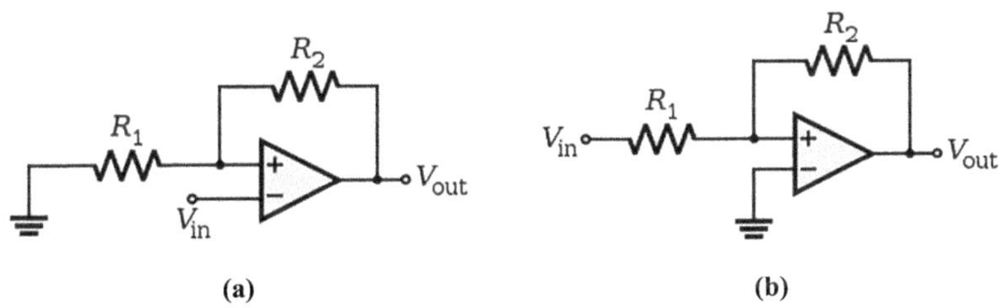

Figure 5.33. Schmitt trigger circuits: (a) with an inverting input. This [Op-Amp Inverting Schmitt Trigger] image has been obtained by the author from the Wikimedia website, where it is stated to have been released into the public domain. It is included within this book on that basis. (b) With a non-inverting input. This [Op-Amp Integrating Amplifier] image has been obtained by the author from the Wikimedia website, where it is stated to have been released into the public domain. It is included within this book on that basis.

In most Op-Amp circuits, the output voltage cannot quite reach the power supply rails. This is because there are small voltage offsets within the Op-Amp's internal circuitry. In addition, the offsets are often different for the positive and negative outputs: $V_{\mathrm{OL}} = V^- + \Delta V_{\mathrm{L}}$, $V_{\mathrm{OH}} = V^+ - \Delta V_{\mathrm{H}}$, where V^- and V^+ are the rail voltages. For example, if the rail voltages are $V^- = -9$ V and $V^+ = 9$ V, the output voltage can swing between $V_{\mathrm{OL}} = -8$ V and $V_{\mathrm{OH}} = 8$ V. These output limits are typically closer to the rail voltages in modern Op-Amps, especially those designed to operate efficiently with lower bias voltages.

The Schmitt trigger circuit with an inverting input, shown in figure 5.33(a), is relatively straightforward to analyze. Since the output voltage switches between two saturation levels (V_{OL} and V_{OH}), the threshold voltages (V_{TL} and V_{TH}) at the non-inverting input—set by the voltage divider—can be easily determined:

$$V_{TL} = \frac{R_1}{R_1 + R_2} V_{OL}, \quad V_{TH} = \frac{R_1}{R_1 + R_2} V_{OH}. \tag{5.30}$$

These voltages define the threshold levels that trigger output transitions when the input signal crosses them. Figure 5.34 shows the simulation results for a Schmitt trigger circuit, where the input voltage is swept between −5 and +5 V. Due to the significant offset voltage of this type of Op-Amp, the limits of the output voltage are −4 V and +4 V. With a voltage divider consisting of two equal resistors, the resulting threshold voltages are −2 V and +2 V, respectively.

Figure 5.34(b) shows the simulation results for both forward and backward sweeps. During the forward sweep, the input signal starts at –5 V, placing the output at its upper limit of +4 V. Under this condition, the threshold voltage is +2 V. When the input voltage rises and crosses this threshold, the output switches to −4 V. Conversely, during the backward sweep, the input starts at +5 V, causing the output to remain at its lower limit of –4 V. In this case, the threshold voltage is −2 V. When the input voltage drops below this value, the output switches back to +4 V.

The two threshold voltages can be adjusted by connecting a DC voltage source (V_2) to the left side of the feedback network. This causes the threshold levels to shift according to the equations shown below. The superposition principle can be applied in the derivation, treating V_2 and the Op-Amp's output voltage as two independent voltage sources:

$$\begin{aligned} V_{TL,S} &= V_{TL} + \frac{R_2}{R_1 + R_2} V_2 \\ V_{TH,S} &= V_{TH} + \frac{R_2}{R_1 + R_2} V_2. \end{aligned} \tag{5.31}$$

Schmitt trigger circuits have a wide range of applications, particularly in converting analog signals into clean digital signals. One of their key advantages is the inherently hysteretic behavior, which introduces two distinct threshold voltages. This

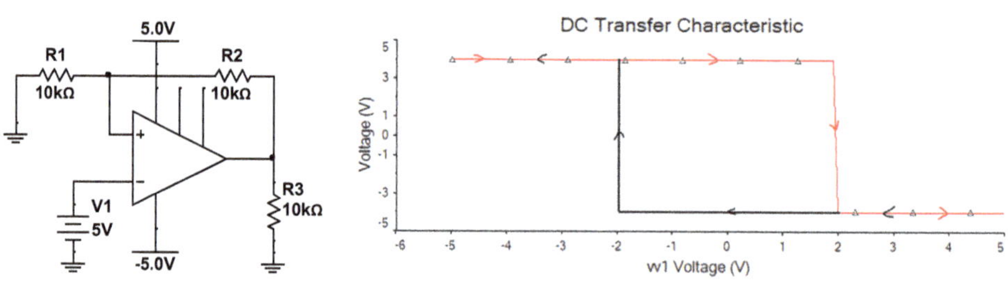

(a) (b)

Figure 5.34. Schmitt trigger with an inverting input: (a) circuit, (b) simulation using a DC sweep.

effectively filters out noise and prevents false triggering when the input signal fluctuates near a single threshold level. As a result, Schmitt triggers are ideal for signal conditioning in noisy environments.

Figure 5.35 illustrates an example where a sine wave is converted into a square wave using a Schmitt trigger. An ideal Op-Amp is used for the simulation. Even when the input sine wave is superimposed upon noise, the output remains a clean, well-defined square wave. This makes Schmitt triggers especially useful in waveform shaping, digital interfacing, and applications requiring noise immunity.

If the feedback network (R1 and R2) is removed from the circuit shown in figure 5.35(a), and the non-inverting input is instead connected directly to a fixed DC voltage source of 2.5 V, the circuit behavior changes drastically, as illustrated in figure 5.36. In this configuration, the Op-Amp functions as a basic comparator without hysteresis. When the input sine wave hovers near the reference voltage, even small amounts of noise superimposed on the signal can cause rapid and erratic switching between the positive and negative supply rails. This phenomenon, known as output chatter, leads to an unstable digital output and can severely compromise the accuracy of analog-to-digital conversion.

In practical systems, such noise-induced fluctuations are unavoidable, making this setup highly susceptible to signal integrity issues. The Schmitt trigger configuration overcomes this limitation by introducing hysteresis through positive feedback, thereby establishing two distinct threshold voltages—one for rising and one

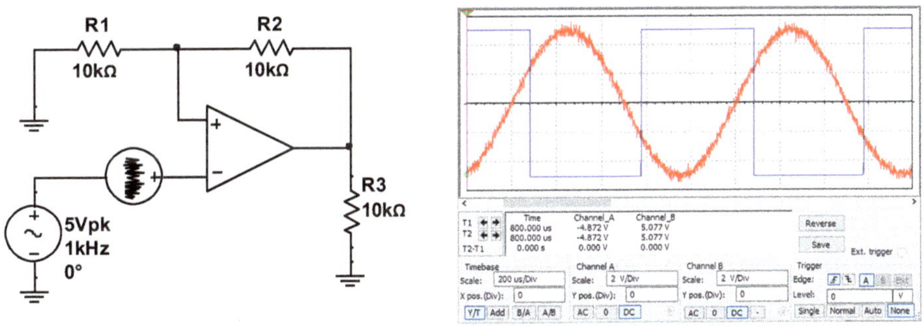

(a) (b)

Figure 5.35. Analog-to-digital conversion: (a) circuit, (b) simulation results.

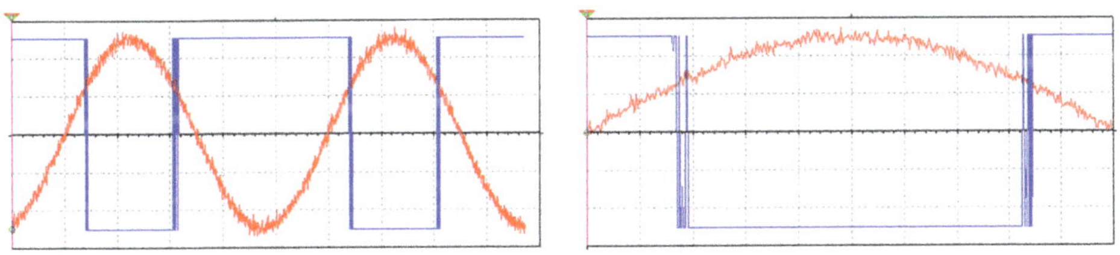

(a) (b)

Figure 5.36. Digitizing with a fixed reference voltage: (a) unmagnified, (b) magnified.

for falling signals—which effectively suppress unwanted transitions due to noise and ensure clean and stable digital output.

In addition to the Schmitt trigger circuit with an inverting input, the circuit with a non-inverting configuration works in a similar way. Since the input signal is connected to a resistor, the input resistance is much lower, which is less desirable compared with the inverting input configuration. Figure 5.37 shows the circuit and the simulation results obtained using a DC sweep.

An intuitive seesaw model was used to analyze the inverting amplifier, and it can also be applied here to find the threshold voltages. However, the *virtual short* condition applies only at the threshold voltage; in other words, the voltages at these two input nodes are different in most situations. Given this condition, the following equations can also be derived:

$$V_{TL} = -\frac{R_1}{R_2}V_{OH}, \quad V_{TH} = -\frac{R_1}{R_2}V_{OL}. \tag{5.32}$$

When a DC voltage is connected to the inverting input of the Op-Amp, the threshold voltages are shifted accordingly. In this scenario, the intuitive hydraulic crane analogy can be used to visualize the behavior and derive the new threshold levels. However, its orientation is flipped: the Op-Amp's output is treated as the ground, while the input node on the left side of the feedback network represents the tip of the hydraulic crane. As a result, the DC voltage applied to the inverting input is effectively amplified through the feedback network, influencing the switching thresholds:

$$\begin{aligned} V_{TL,S} &= V_{TL} + \frac{R_1 + R_2}{R_2}V_2 \\ V_{TH,S} &= V_{TH} + \frac{R_1 + R_2}{R_2}V_2. \end{aligned} \tag{5.33}$$

Figure 5.38 shows an example of converting a sine wave into a square wave using a Schmitt trigger configured with a non-inverting input; an ideal Op-Amp is used in the simulation. If the inverting input is grounded, the threshold voltages are −2.5 V and +2.5 V. However, applying a 1 V DC offset voltage shifts the threshold levels by

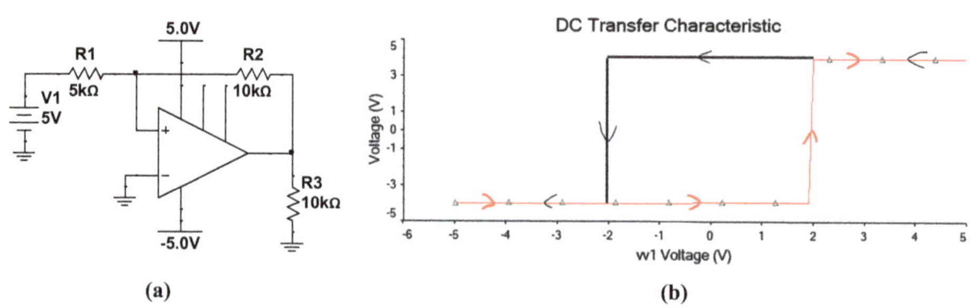

Figure 5.37. Schmitt trigger with a non-inverting input: (a) circuit, (b) simulation with DC sweep.

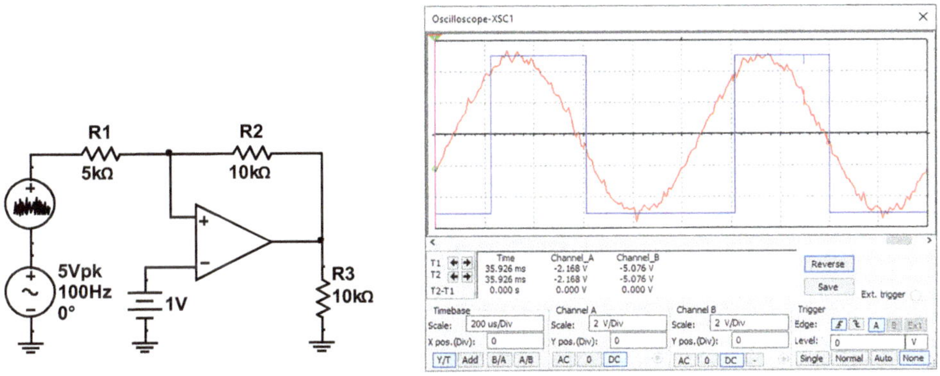

(a) (b)

Figure 5.38. Analog-to-digital conversion: (a) circuit, (b) simulation results.

1.5 V, resulting in new threshold voltages of −1 V and 4 V, respectively. The simulation results in figure 5.38(b) illustrate this effect, showing a square-wave output with an uneven duty cycle due to the asymmetrical threshold levels.

In addition to Schmitt triggers, circuits employing positive feedback play a vital role in many analog and mixed-signal systems. Unlike negative feedback, which stabilizes a system by opposing changes, positive feedback reinforces changes, resulting in bistable, regenerative, or oscillatory behavior. In the early days of electronics, particularly in circuits using vacuum tubes, positive feedback was commonly employed to enhance amplifier gain.

IOP Publishing

Essential Microelectronic Circuits (Second Edition)
A student's guide

Yumin Zhang

Chapter 6

Oscillator circuits

Oscillator circuits are fundamental building blocks in analog and digital electronic systems. In contrast to amplifiers that depend on external input signals, oscillators can generate continuous, periodic waveforms—such as sine waves, square waves, or triangular waves—without any input. These circuits convert DC power into an AC signal at a desired frequency. Oscillators are essential in a wide range of applications, including signal generation in analog circuits and clock generation in digital systems. Broadly, oscillator circuits can be categorized into two main types:

- **sinusoidal oscillators**, which generate smooth, continuous sine waves;
- **multivibrators**, which produce abrupt transitions in waveforms, such as square waves.

This chapter introduces the fundamental principles and practical implementations of oscillator circuits, which are essential for generating periodic signals in electronic systems. Section 6.1 begins with the concept of positive feedback, establishing the basic criterion for oscillation. Sections 6.2–6.5 explore oscillator circuits using RC and LC feedback networks, including classic configurations such as the Wien-bridge, Hartley, and Colpitts oscillators. Sections 6.6 and 6.7 delve into negative-resistance oscillators, explaining their operating mechanisms and presenting transistor-based designs. Section 6.8 investigates coupled oscillator circuits, analyzing mode selection and the effect of initial conditions. Finally, section 6.9 discusses multivibrator circuits, including astable, monostable, and bistable configurations, which produce non-sinusoidal waveforms and serve as timers or memory elements. Throughout the chapter, simulation results and waveform analyses are provided to support theoretical discussions and enhance practical understanding.

doi:10.1088/978-0-7503-5512-4ch6

6.1 Oscillators with positive feedback

In the previous chapter, systems with negative feedback were discussed. In this chapter, we turn our attention to systems with positive feedback, which form the basis of oscillator circuits that generate sine waves. As shown in figure 6.1(a), the block diagram of a typical positive feedback system includes an amplifier and a feedback network. The amplifier is usually designed with broadband characteristics to provide sufficient gain across a wide range of frequencies. The feedback path, characterized by a frequency-dependent gain factor $\beta(\omega)$, plays a crucial role in determining the oscillation frequency. When the loop gain satisfies the *Barkhausen criterion* at a specific frequency, $A\beta(\omega_o) = 1$, the system can sustain stable oscillations without any external input signal. This principle is analogous to the operation of a laser, in which light bounces back and forth between two mirrors inside a resonant cavity with unity loop gain, as illustrated in figure 6.1(b).

Using an approach similar to the analysis of the negative feedback system, the relationships between the four variables in the block diagram can be set up for the three devices in the system as follows:

$$\begin{aligned} x_d &= x_i + x_f \\ x_o &= Ax_d \\ x_f &= \beta x_o. \end{aligned} \tag{6.1}$$

Replacing the internal variables (x_d, x_f) with the external variables (x_i, x_o) allows the feedback gain to be derived:

$$A_f = \frac{x_o}{x_i} = \frac{A}{1 - A\beta}. \tag{6.2}$$

In oscillator circuits, the traditional feedback gain formula used for amplifiers—where the system is driven by an external input—is not directly applicable, since oscillators operate without any external input signal, $x_i = 0$. Instead, the concept of ***loop gain***, denoted by $L(s) = A(s)\beta(s)$, becomes more relevant. Consider a hypothetical signal $x_d(0) = x_f(0)$ that appears at the input of the amplifier. First, it is boosted by the amplifier, $x_o(0) = A(s)x_f(0)$. Then, the signal passes through the feedback network, resulting in a returned signal, $x_f(1) = \beta(s)x_o(0) = A(s)\beta(s)x_f(0)$. Thus, the loop gain characterizes the total gain in one complete round trip through the loop. This regenerative process is analogous to the behavior of light circulating

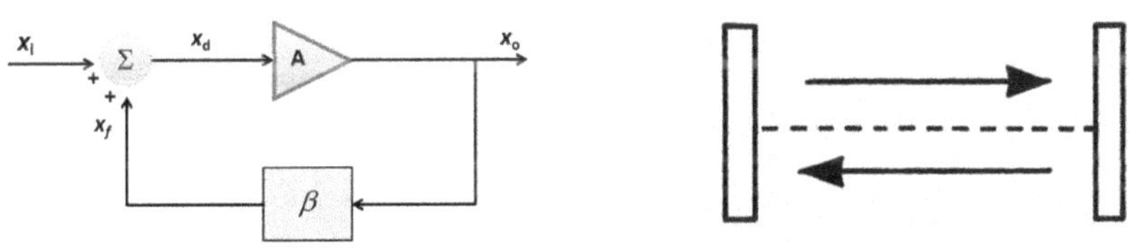

(a) (b)

Figure 6.1. Systems with positive feedback: (a) block diagram, (b) laser cavity. Part (b) created with GPT-4.0, OpenAI.

in a laser cavity, where the intensity builds up through repeated amplification. In oscillator circuits, this leads to self-sustaining oscillation, known as a ***limit cycle*** in nonlinear dynamics—a phenomenon that cannot be fully captured using only linear models.

> A **limit cycle** in a nonlinear system is a closed, isolated trajectory in the system's phase space that represents a stable, self-sustained oscillation. Unlike linear systems, nonlinear systems with limit cycles naturally evolve toward this repeating behavior regardless of initial conditions (within a certain region). Once the system reaches the limit cycle, it continues to oscillate with a fixed amplitude and period. Limit cycles are fundamental in explaining phenomena such as sustained biological rhythms and stable electronic oscillations in nonlinear circuits.

In practice, the loop gain cannot remain constant. If the gain were fixed at exactly unity, the oscillation would not be able to start. To enable the onset and maintenance of oscillation, the loop gain must be a function of the signal amplitude. When the signal is very weak, the loop gain should be greater than one to allow the signal to grow. As the signal amplitude increases, nonlinearities within the circuit reduce the loop gain, preventing runaway growth. Ultimately, the system reaches a steady state where the loop gain is maintained at unity—this balance defines the boundary between the growth and decay of the signal. This dynamic equilibrium point, where the oscillation stabilizes at a constant amplitude and frequency, is governed by the *Barkhausen criterion*, as shown in the following equation:

$$A(\omega_o)\beta(\omega_o) = 1 \rightarrow |A| \cdot |\beta| = 1, \quad \angle A + \angle \beta = n \cdot 360^\circ. \tag{6.3}$$

Compared to amplifiers, oscillator circuits may seem more mysterious because they appear to generate a periodic signal without any external input. However, from a physical perspective, the origin of oscillation can be attributed to the inherent electrical noise present in all electronic circuits. This noise consists of random fluctuations with components that span a wide range of frequencies, effectively acting as a weak broadband input signal.

If the amplifier in an oscillator circuit has a relatively flat gain over the relevant frequency range, it amplifies all frequency components nearly equally. In this case, it is the feedback network that plays the critical role: it determines the specific frequency at which the overall loop gain and phase conditions are satisfied—typically based on the *Barkhausen criterion*. Thus, while it may seem as though oscillators 'create' signals from nothing, they actually exploit noise and carefully engineered feedback to initiate and maintain sinusoidal waveforms.

Figure 6.2(a) illustrates the nonlinearity effect of an amplifier, where the gain decreases when the amplitude or power of the signal becomes stronger. Initially, the gain is higher than the loss from the feedback network, so the signal grows stronger and stronger. In this diagram, the sign of the loss is flipped; for example, if the loss is

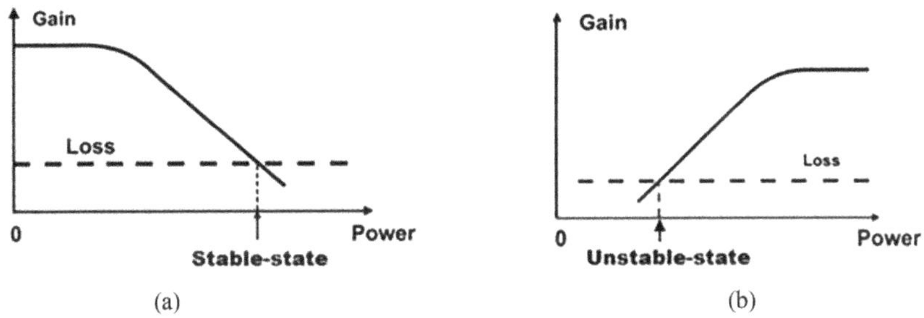

Figure 6.2. Gain decrease with signal power: (a) stable case, (b) unstable case. Created with GPT-4.0, OpenAI.

−10 dB, the dashed line here is at +10 dB. As the signal grows, the gain of the amplifier decreases until it is equal to the flipped loss, which is the same as the Barkhausen criterion:

$$\text{Gain(dB)} = -\text{Loss(dB)} \Rightarrow \text{Gain(dB)} + \text{Loss(dB)} = 0\text{(dB)} \tag{6.4}$$

It should be noted that figure 6.2(a) illustrates the case of stable oscillation, which functions as an *attractor* in the system once the Barkhausen criterion is satisfied. In practical electronic circuits, noise and perturbations are always present and can momentarily alter the signal amplitude. When the amplitude increases slightly, it moves to the right side of the steady-state point in the diagram, making the gain fall below the loss, which causes the amplitude to decrease. Conversely, when the amplitude becomes lower, it moves to the left side of the steady-state point in the diagram, the gain exceeds the loss, and then the signal becomes stronger. As a result, the system continuously adjusts itself, allowing small fluctuations but keeping the oscillation bounded and sustained over time. This behavior is characteristic of a *limit cycle* in nonlinear dynamic systems.

Figure 6.2(b) illustrates an unstable oscillation condition, which can be analyzed similarly to the stable case. In this scenario, the operating point lies at an unstable equilibrium, where small deviations in signal amplitude lead to further divergence. If the signal becomes slightly stronger, it moves to the right of the intersection point, where the gain exceeds the loss, causing the amplitude to grow uncontrollably. Conversely, if the signal weakens slightly, it moves to the left of the crossing point, where the gain is below the loss, and the amplitude decays toward zero. This positive feedback loop results in either signal explosion or extinction, rather than stabilization. Fortunately, most well-designed amplifier circuits exhibit automatic amplitude stabilization behavior, as illustrated in figure 6.2(a), which ensures that stable, self-sustained oscillation is achievable in practical systems.

6.2 Oscillator circuits with RC feedback networks I

6.2.1 Wien-bridge oscillator circuit

The Wien-bridge oscillator is a classic example of a system with positive feedback, as shown in figure 6.3. It uses a frequency-dependent circuit in the feedback path to favor a single frequency, allowing a clean and stable sinusoidal oscillation to emerge.

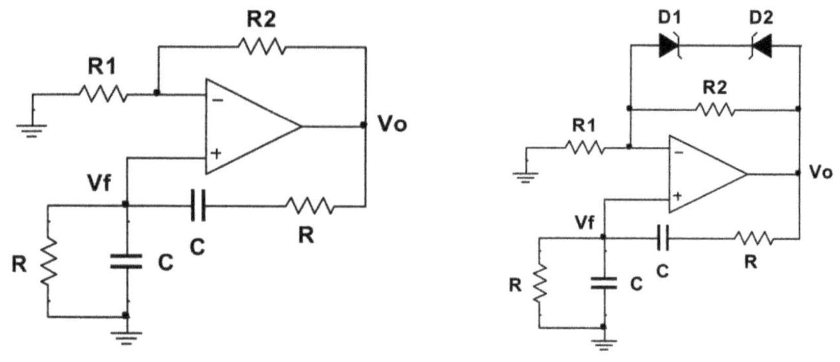

Figure 6.3. Wien-bridge oscillator circuits: (a) simplified version, (b) practical version.

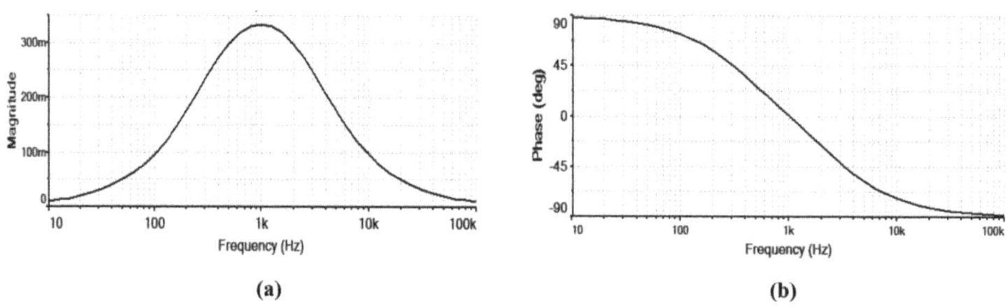

Figure 6.4. Frequency response of a lead–lag network: (a) magnitude, (b) phase.

This example effectively demonstrates how collaboration between the amplifier and the feedback network shapes the behavior of oscillator circuits.

Figure 6.3(a) shows the simplified Wien-bridge oscillator circuit. The inverting input of the Op-Amp is connected to a feedback path consisting of two resistors (R_1 and R_2), which determine the voltage gain: $A = 1 + R_2/R_1$. The non-inverting input of the Op-Amp is connected to the second feedback network consisting of resistors and capacitors, which form the so-called *lead–lag network* that determines the feedback factor β. This circuit is similar to a voltage divider composed of a series component and a parallel component:

$$\beta(s) = \frac{\widetilde{V}_{\mathrm{f}}}{\widetilde{V}_{\mathrm{o}}} = \frac{Z_{\mathrm{P}}}{Z_{\mathrm{S}} + Z_{\mathrm{P}}} = \frac{1}{1 + Z_{\mathrm{S}} Y_{\mathrm{P}}} = \frac{1}{1 + (R + 1/sC)(1/R + sC)} = \frac{1}{3 + j(\omega/\omega_{\mathrm{o}} - \omega_{\mathrm{o}}/\omega)}. \tag{6.5}$$

In this equation, Z_{S} and Z_{P} are the impedances of the series and parallel RC circuits, respectively, and $\omega_{\mathrm{o}} = 1/RC$. The frequency response of this feedback factor is shown in figure 6.4. At the resonant frequency, the feedback factor reaches its peak value, $\beta(\omega_{\mathrm{o}}) = 1/3$, and the phase shift is zero. Following the Barkhausen criterion, the voltage gain of the amplifier should be exactly three at the resonant frequency. However, to initiate oscillation from the inherent noise in the circuit, the gain must initially be slightly greater than three. Once the oscillation amplitude increases, the gain must be automatically reduced to stabilize the signal and prevent distortion.

As shown in figure 6.3(b), automatic gain control is achieved using two Zener diodes in the amplifier's negative feedback path at the top. When the output signal is small, the Zener diodes are nonconductive, and the circuit operates with high gain. As the signal grows and exceeds the Zener breakdown voltage, the diodes conduct and effectively limit the gain, helping stabilize the output at a constant amplitude. This mechanism enables the Wien-bridge oscillator to produce a clean sinusoidal waveform without external intervention.

6.2.2 Waveform and spectrum

Figure 6.5(a) shows a simulated waveform from the simplified Wien-bridge oscillator circuit presented in figure 6.3(a) with the following component values: R1 = 10 kΩ, R2 = 22 kΩ, R = 10 kΩ, and C = 10 nF. The resonant frequency is $f_o = 1.59$ kHz, and the voltage gain is 3.2 V/V, slightly above the critical threshold required for oscillation. As a result, the oscillation starts quickly; however, the amplitude continues to increase unchecked until the output signal reaches the supply voltage limits of the Op-Amp, leading to clipping and waveform distortion.

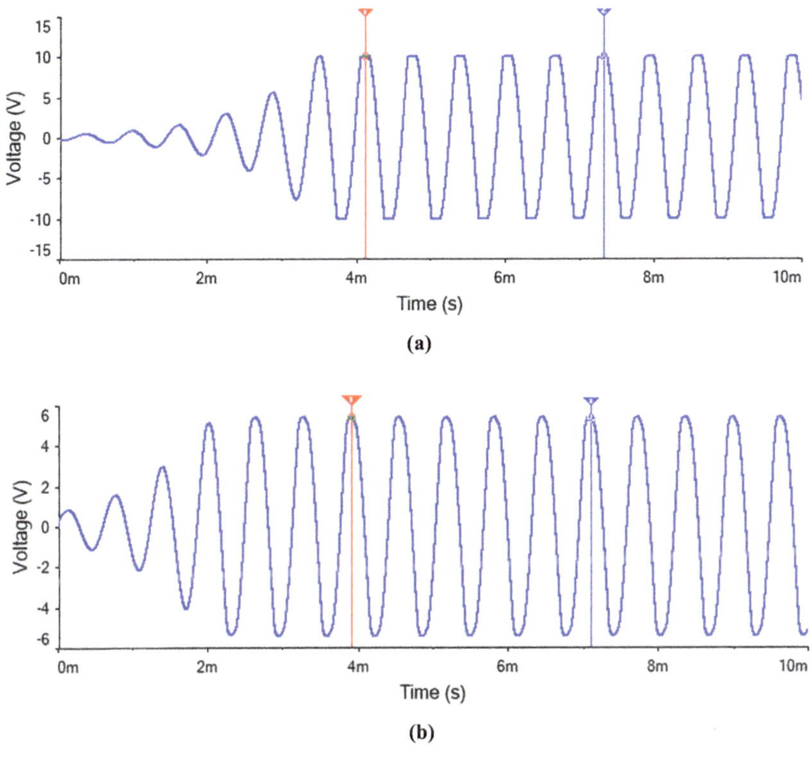

Figure 6.5. Waveforms produced by a Wien-bridge oscillator: (a) without Zener diodes, (b) with Zener diodes.

A few tricks are used in simulating oscillator circuits. First, at least one capacitor in the circuit should be assigned a nonzero initial voltage. Second, in the 'Transient Analysis' settings, it is crucial to select 'User-defined' as the initial conditions. Third, oscillators may take some time to build up from the noise level to a steady-state sine wave.

In contrast, figure 6.5(b) shows the simulated waveform from the practical Wien-bridge oscillator circuit presented in figure 6.3(b), which incorporates amplitude stabilization using Zener diodes. In this configuration, the output amplitude is automatically regulated, preventing excessive growth. As a result, the waveform is significantly improved, maintaining a stable sinusoidal shape without clipping. This demonstrates the importance of implementing nonlinear amplitude control in practical oscillator designs to ensure reliable operation. By measuring the period of the waveform, the frequency is determined to be $f_o = 1.57$ kHz, which agrees very well with the theoretical result. In addition, this frequency can also be measured with a voltage probe, as shown in figure 6.7(a).

Multisim offers a powerful virtual tool for signal analysis: the **spectrum analyzer**, which emulates the functionality of its physical counterpart used in electronic testing. Unlike an oscilloscope that displays signals in the time domain, a spectrum analyzer presents signals in the frequency domain, showing how the signal power is distributed across different frequencies. This capability is especially valuable in oscillator circuit analysis, where it helps verify the dominant frequency of oscillation, measure spectral purity, and detect unwanted harmonics or spurious signals. Additionally, the spectrum analyzer aids in evaluating the signal bandwidth and identifying the presence of noise or distortion components. By providing a clear view of the signal's frequency content, it offers deeper insight into the behavior and performance of oscillator circuits and other signal-generating systems.

Figure 6.6(a) shows the output spectrum obtained from the virtual spectrum analyzer connected to the output node of the Wien-bridge oscillator circuit. In addition to the dominant peak at the fundamental frequency, two smaller harmonic peaks are observed. The first is the third harmonic at approximately 4.75 kHz, and the second is the fifth harmonic at about 7.98 kHz. The third harmonic peak is approximately 18 dB below the fundamental, suggesting a moderate level of harmonic distortion. Figure 6.6(b) displays the result from the virtual distortion analyzer, showing a total harmonic distortion (THD) of 7.3%, which confirms that a significant amount of distortion is present in the waveform.

Notably, even-order harmonics are absent in the output spectrum, which suggests that the waveform produced by the Wien-bridge oscillator exhibits a high degree of symmetry. For example, regular square waves and triangular waves also have this symmetry in the waveform in each half-cycle. The Fourier series of such waveforms have either odd or even components, depending on the location of the origin on the time axis. By contrast, asymmetric waveforms—such as the sawtooth wave—lack this symmetry and consequently contain both even and odd harmonics.

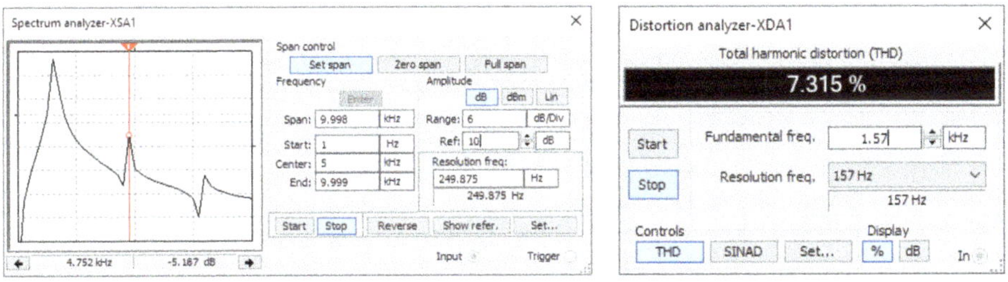

(a) (b)

Figure 6.6. Spectrum analysis: (a) spectrum analyzer, (b) distortion analyzer.

6.2.3 Oscillators with filters

To improve the purity of the output waveform, a first-order low-pass filter (LPF) is incorporated, as depicted in figure 6.7. This filter is designed so that its cutoff frequency is precisely set to the oscillation frequency. By attenuating unwanted higher-order harmonics generated by the oscillator's nonlinearities, the LPF can effectively clean up the waveform. Simulation results demonstrate a significant reduction of distortion, with the THD notably reduced from 7.3% to 2.4%. This highlights the critical role of LPFs in removing higher harmonics from the outputs of oscillator circuits.

Even with the first-order LPF, a THD of 2.4% is still observed, which is considered quite high for applications requiring a very clean sinusoidal waveform. To achieve further waveform purity, a second-order Sallen–Key LPF was implemented, as illustrated in figure 6.8(a). This higher-order filter offers a steeper roll-off characteristic, providing more aggressive attenuation of unwanted harmonics.

The simulation results confirm its effectiveness: the THD was dramatically reduced from the initial 7.3% all the way down to an impressive 1.0%. Figure 6.8(b) visually highlights this significant improvement in the waveforms: the top trace displays the original unfiltered signal, while the bottom trace clearly shows the output signal after passing through the second-order LPF. The visual contrast unequivocally demonstrates the superior performance of this second-order filter in enhancing signal quality.

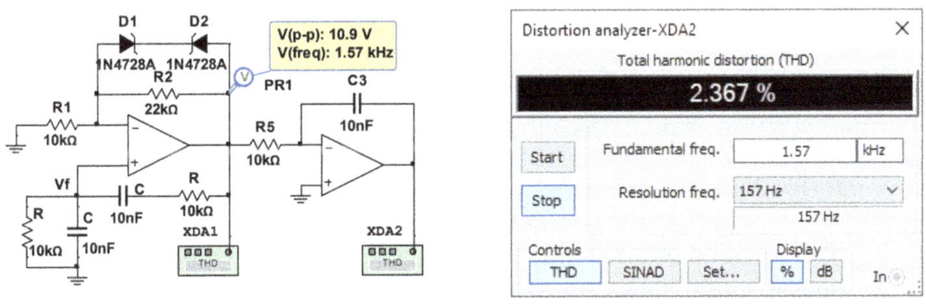

(a) (b)

Figure 6.7. Distortion reduction by first-order filtering: (a) circuit, (b) THD value.

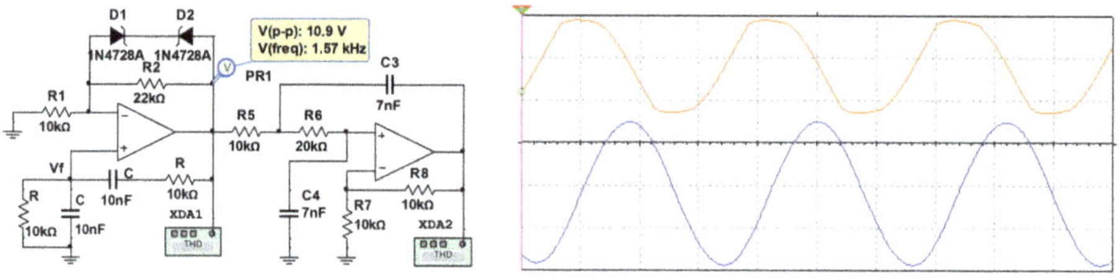

(a) (b)

Figure 6.8. Reducing distortion via second-order filtering: (a) circuit, (b) waveforms.

William Hewlett—the co-founder of Hewlett-Packard Company (HP)—investigated the Wien-bridge oscillator in his master's thesis at Stanford University in 1939. At that time, semiconductor diodes and transistors were not yet invented, so he used vacuum tubes in his circuit. The nonlinear gain behavior was implemented using an incandescent lamp (filament bulb) in place of resistor R_1 in the oscillator circuit. The filament's resistance increased with temperature, providing a form of automatic gain control: when the output amplitude rose, the filament heated up, its resistance increased, and the loop gain decreased, thereby stabilizing the oscillation amplitude. This innovative use of a temperature-dependent resistor enabled the circuit to produce low-distortion sine waves without the need for manual adjustment. Hewlett's design was implemented in HP's first commercial product—the HP200A audio oscillator—which became a key tool in audio and electronic testing.

> Although the Wien-bridge oscillator is no longer widely used in modern designs due to the availability of more precise and stable oscillator circuits, it remains an important historical milestone in analog electronics and is still used in educational and low-frequency analog applications.

6.3 Oscillator circuits with RC feedback networks II

The Barkhausen criterion, expressed in equation (6.3), $A(\omega_o)\beta(\omega_o) = 1$, is a complex equation comprising two real components: one for magnitude and one for phase. While the magnitude condition—which requires that the loop gain equals unity—is often emphasized in analyzing frequency-selective feedback networks, the phase condition is equally critical for sustained oscillation. This can be clearly illustrated using the RC phase-shift oscillator circuit shown in figure 6.9(a). The feedback network, located on the right-hand side of the Op-Amp, consists of three cascaded RC high-pass filter (HPF) sections. For the circuit to oscillate, the total phase shift introduced by the feedback network must be 180°, so that when combined with the

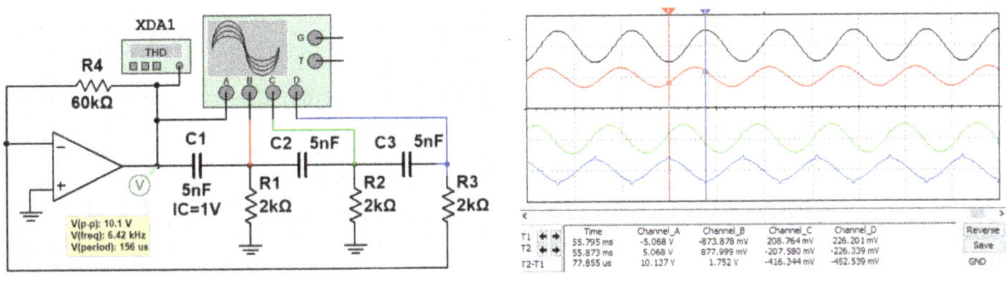

(a) (b)

Figure 6.9. Phase-shift oscillator with HPFs: (a) circuit, (b) waveforms.

additional 180° phase shift provided by the inverting Op-Amp, the total phase shift around the loop is 360°, satisfying the phase requirement of the Barkhausen criterion.

Figure 6.9(b) displays the resulting waveforms. The cursors in the figure measure the time delay between the output and feedback signal, which corresponds to half the oscillation period—consistent with a 180° phase shift. Additionally, the output waveform of the Op-Amp is nearly sinusoidal, with a THD of approximately 0.56%, indicating good waveform purity.

At first glance, the feedback network on the right-hand side of the Op-Amp may appear to be simply three cascaded RC HPF sections, each contributing a 60° phase shift, thereby producing the required 180° total phase shift. However, the actual behavior is more complex due to the **loading effect**—the influence of each stage on the impedance of the previous stages. In deriving the transfer function of a first-order RC HPF, the voltage divider principle is typically used, where the circuit is partitioned into two impedance components, Z_1 and Z_2. For instance, when analyzing the section across capacitor C_1, Z_1 corresponds to the impedance of C_1, but Z_2 is not simply R_1; instead, it should include the impedance of the following two RC stages. As a result, the transfer function of the overall feedback network becomes more intricate. Nevertheless, through proper analysis, the transfer function can be derived:

$$H(j\omega) = \frac{-j(\omega/\omega_c)^3}{[1 - 6(\omega/\omega_c)^2] + j[5(\omega/\omega_c) - (\omega/\omega_c)^3]}. \tag{6.6}$$

In this equation, $\omega_c = 1/RC$. To satisfy the condition for a phase shift of 180°, the first term in the denominator of the transfer function must vanish. This leads to the formula for the resonant frequency: $\omega = \omega_c/\sqrt{6}$ and $f = 1/(2\pi\sqrt{6}RC)$, which is equal to 6.50 kHz. This theoretical value compares well with the simulation result of 6.42 kHz shown in figure 6.9(a), demonstrating good agreement and validating the analytical model despite the presence of circuit imperfections.

At the resonant frequency, $\omega = \omega_c/\sqrt{6}$, the magnitude of the transfer function given in equation (6.6) is $-1/29$. According to the Barkhausen magnitude criterion, the inverting amplifier must provide a gain of at least 29 V/V to overcome the attenuation introduced by the RC feedback network and sustain oscillation. In the circuit shown in figure 6.9(a), the amplifier gain is set to −30 V/V, slightly exceeding the threshold to ensure reliable startup.

Since the circuit lacks Zener diodes or any other amplitude-limiting components, the oscillation amplitude continues to grow until it reaches the supply rail limits of the Op-Amp, which are ±5 V in this design. This saturation behavior is clearly reflected in the voltage probe measurements shown in figure 6.9(a). As the signal travels through the cascading RC sections, it experiences progressive attenuation. Consequently, the waveforms in figure 6.9(b) are displayed using different vertical scales—specifically, 10, 5, 1, and 0.5 V/div, respectively—to accommodate the decreasing amplitudes at each stage of the feedback network.

Since a three-stage RC HPF network can provide a total phase shift of 180°, a natural question arises: would the oscillator still function if the HPFs were replaced with LPFs? At first, this seems counterintuitive, since LPFs introduce negative phase shifts. However, a phase shift of −180° is equivalent to +180°, so the Barkhausen phase condition can still be satisfied. Figure 6.10(a) shows such a circuit using three cascaded RC LPF stages. An additional resistor, R_4, is included in the inverting amplifier to set the gain appropriately.

Although the component values (resistors and capacitors) are the same as those in the HPF-based oscillator circuit in figure 6.9(a), the resulting resonant frequency is significantly higher—about 41.7 kHz. Figure 6.10(b) presents the corresponding waveforms, using the same vertical scale scheme as in figure 6.9(b). The top trace shows severe saturation and distortion at the Op-Amp output, but the waveform quality improves as the signal travels through the cascading LPF stages. The bottom trace exhibits a clean sinusoidal waveform with a THD of just 0.62%. By comparing the results of figure 6.10(b) and figure 6.9(b), it becomes evident that LPF stages tend to enhance waveform quality by attenuating higher-order harmonic components. In contrast, HPF stages introduce more distortion and degrade signal integrity, as they attenuate the fundamental frequency while allowing higher-frequency harmonics to remain relatively strong.

In designing an RC phase-shift oscillator, a critical design parameter is the value of the feedback resistor R_5, which sets the gain of the inverting amplifier. According to the Barkhausen magnitude criterion, this gain must be sufficiently large to compensate for the attenuation introduced by the RC feedback network. While the

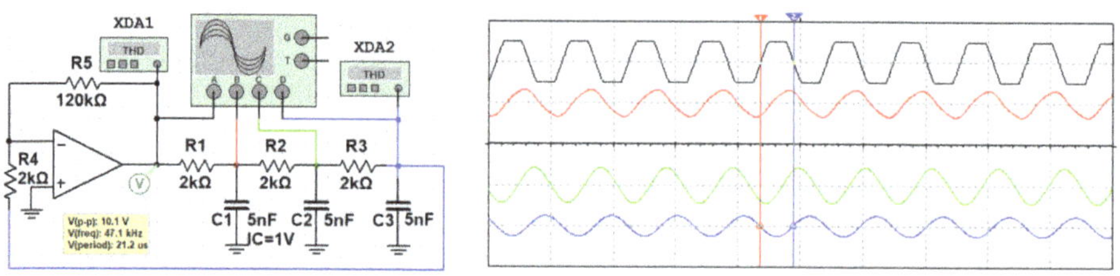

(a) (b)

Figure 6.10. Phase-shift oscillator with LPFs: (a) circuit, (b) waveforms.

transfer function of the three-stage RC network can be derived analytically, practical insight is often gained through circuit simulation.

Figure 6.11 presents Bode plots of the RC feedback network, including resistor R_4 from the oscillator circuit in figure 6.10(a). The phase response is shown in the upper plot of figure 6.11(b), where the cursor marks the frequency at which the total phase shift reaches $-180°$, identified as $f = 50.5$ kHz. The corresponding magnitude response in the lower plot of figure 6.11(b) indicates a gain of approximately -35 dB, or -0.0178 V/V. This result is consistent with the value obtained directly from the voltage probe in figure 6.11(a), which indicates a gain of -0.0176 V/V. Using this feedback gain, the required gain of the inverting amplifier can be calculated to satisfy the Barkhausen magnitude condition: $|A_V| > 57\ \mathrm{V/V}$. To ensure reliable startup, a slightly higher gain of –60 V/V is selected in the oscillator circuit shown in figure 6.10(a), providing a safety margin above the minimum required value.

A noticeable discrepancy exists between the resonant frequency (47.1 kHz) observed in the simulation of the oscillator circuit in figure 6.10(a), and the –180° phase-shift frequency (50.5 kHz) identified in the Bode plot of figure 6.11(b). This naturally raises the question: could a similar deviation affect the required voltage gain obtained from the simulation results in figure 6.11? To investigate this, a trial-and-error approach was used to adjust the value of the feedback resistor R_5.

The oscillator continues to function reliably when R_5 is reduced to 107 kΩ, and interestingly, the oscillation frequency shifts to 49.2 kHz, which is closer to the result (50.5 kHz) shown in figure 6.11(b). In addition, the distortion of the waveform at the output node of the Op-Amp also decreases significantly, suggesting operation closer to the ideal gain condition. However, when R_5 is further reduced to 106 kΩ, the oscillation disappears, indicating that the loop gain has fallen below the critical threshold. This result implies that the minimum required gain of the inverting amplifier is slightly greater than 53 V/V. Despite the observed discrepancies in frequency and gain, the simulation results in figure 6.11 remain highly useful for guiding the design and tuning of the oscillator circuit.

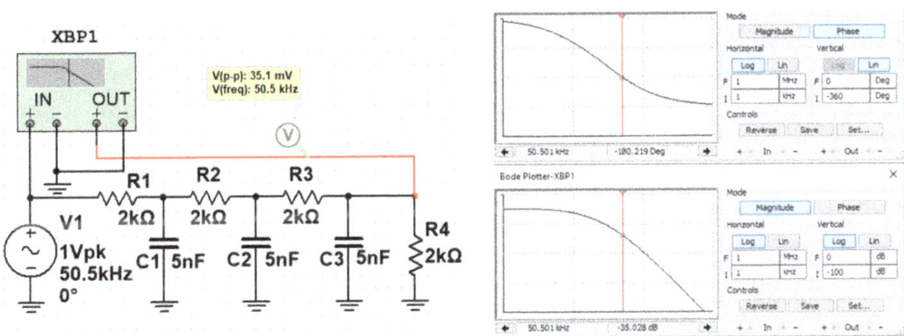

(a) (b)

Figure 6.11. RC LPF network: (a) circuit, (b) Bode plots.

Phase-shift oscillators are widely used in applications that require low-frequency sinusoidal waveforms with stable amplitude and frequency. Due to their simplicity and use of resistors, capacitors, and a single active device (such as an Op-Amp or transistor), they are ideal for audio frequency generation, function generators, and test equipment. In instrumentation systems, they serve as reference signal sources or clock generators. Their predictable frequency characteristics also make them useful in phase-locked loop (PLL) systems and analog signal processing. Additionally, phase-shift oscillators are often used in educational settings to demonstrate fundamental oscillator principles and frequency-domain behavior.

6.4 Oscillator circuits with LC feedback networks I

6.4.1 Analysis of LC networks

The frequency spectra of the output signals produced by Wien-bridge oscillators are typically not very clean, primarily because the RC feedback network lacks strong frequency selectivity, as illustrated in figure 6.4. In other words, the quality factor is too low. The broad bandwidth of the RC network allows multiple harmonics and noise components to persist in the output. To achieve a purer sinusoidal output, alternative feedback networks with higher selectivity are often employed. One such approach is the use of LC resonant circuits, which provide sharper resonance characteristics due to their higher quality factor (Q).

Figure 6.12(a) presents a representative LC feedback network used in oscillator circuits. In this model, parasitic series resistances associated with the inductors are explicitly included to more accurately reflect real-world component behavior. These resistances introduce damping and reduce the quality factor (Q) of the resonant circuit. However, to simplify the theoretical analysis and emphasize the core resonance mechanism, these resistors are neglected in the subsequent derivations.

Historically, the configuration with two inductors originated from a single inductor with a central tap. This approach allowed circuit designers to implement feedback paths and impedance transformations more effectively, particularly in early RF oscillator designs such as the Hartley oscillator. By splitting the inductor, one can control the feedback ratio by adjusting the relative inductance values of the two halves, thereby tuning the oscillation conditions without additional components.

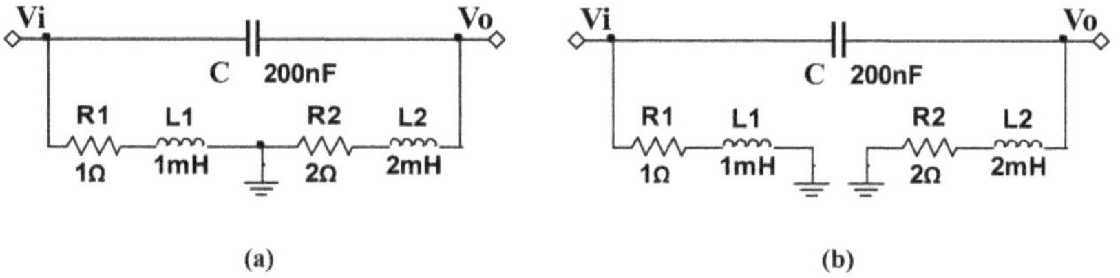

Figure 6.12. LC feedback network with a split inductor: (a) original circuit, (b) transformed circuit.

In the derivation of the transfer function, the original circuit can be transformed into the form shown in 6.12(b), where the inductor L_2 and the capacitor are arranged in series, while the inductor L_1 is not involved. The transfer function of this series LC network can be derived from the formula for a voltage divider:

$$H(\omega) = \frac{V_o}{V_i} = \frac{Z_{L2}}{Z_C + Z_{L2}} = \frac{Z_{L2} Y_C}{1 + Z_{L2} Y_C} = -\frac{\omega^2 L_2 C}{1 - \omega^2 L_2 C} = -\frac{\omega^2/\omega_2^2}{1 - \omega^2/\omega_2^2}. \quad (6.7)$$

In this equation, $\omega_2 = 1/\sqrt{L_2 C}$, which is the series resonant frequency between L_2 and the capacitor. Figure 6.13 shows the Bode plot of this circuit, where a distinct peak in the magnitude response occurs at $f_2 = \omega_2/2\pi$, confirming the resonance condition. This is also reflected in the phase plot, where the phase shift is 90°. Notably, there is a sharp transition in the phase plot near this resonant frequency: the phase is 180° for $\omega < \omega_2$, while it equals 0° for $\omega > \omega_2$. This behavior aligns with the theoretical prediction from equation (6.7), verifying the expected phase characteristics of the LC resonance.

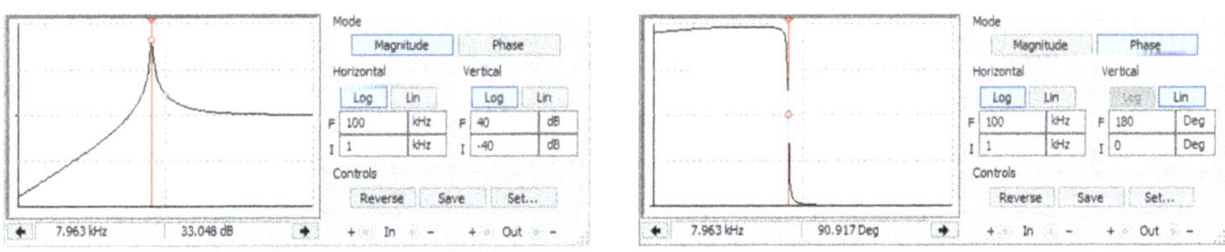

(a) (b)

Figure 6.13. Bode plot of an LC feedback network: (a) magnitude, (b) phase.

If the ground connection between the two inductors is ignored, the resonant frequency—of the parallel LC circuit—can be determined: $\omega_o = 1/\sqrt{(L_1 + L_2)C}$, which is less than the resonant frequency between L_2 and C: $\omega_o < \omega_2$. At this parallel frequency ω_o, the transfer function has a very simple expression:

$$H(\omega_o) = -\frac{\omega_o^2 L_2 C}{1 - \omega_o^2 L_2 C} = -\frac{L_2/(L_1 + L_2)}{1 - L_2/(L_1 + L_2)} = -\frac{L_2}{L_1}. \quad (6.8)$$

Interestingly, the circuit's behavior resembles that of a mechanical seesaw, where the impedances of L_1 and L_2 play roles analogous to lever arms on either side of the pivot. In other words, it can be derived from the bottom branch with the two inductors, though the capacitor is not irrelevant. Moreover, the capacitor can conceptually be divided into two segments, with their impedance ratio matching that of the inductors, as illustrated in figure 6.14(b). If the ground node originally located between the inductors is instead placed at the midpoint between the two capacitors—and the two inductors are combined into a single inductor—the overall transfer function remains unchanged. This demonstrates that the two configurations are functionally equivalent in terms of their frequency response at the parallel resonant frequency:

$$H(\omega_o) = -\frac{Z_2}{Z_1} = -\frac{L_2}{L_1} = -\frac{C_1}{C_2}. \quad (6.9)$$

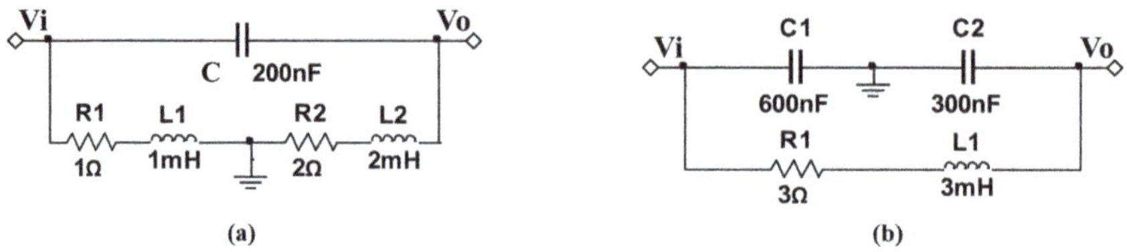

Figure 6.14. LC networks: (a) with a split inductor, (b) with a split capacitor.

6.4.2 Oscillator based on an Op-Amp

At the parallel resonant frequency, equation (6.9) shows that the LC tank circuit provides a 180° phase shift. In addition, the magnitude of the transfer function at this frequency depends on how the inductor is partitioned. When this resonant network is placed within the feedback loop of an inverting amplifier—already contributing another 180° phase shift—the total phase shift around the loop becomes 360°, thus satisfying the phase condition of the Barkhausen criterion. The amplifier also supplies the necessary gain to compensate for energy losses in the LC circuit, thereby meeting the amplitude condition of the criterion as well. Under these conditions, sustained oscillation can occur.

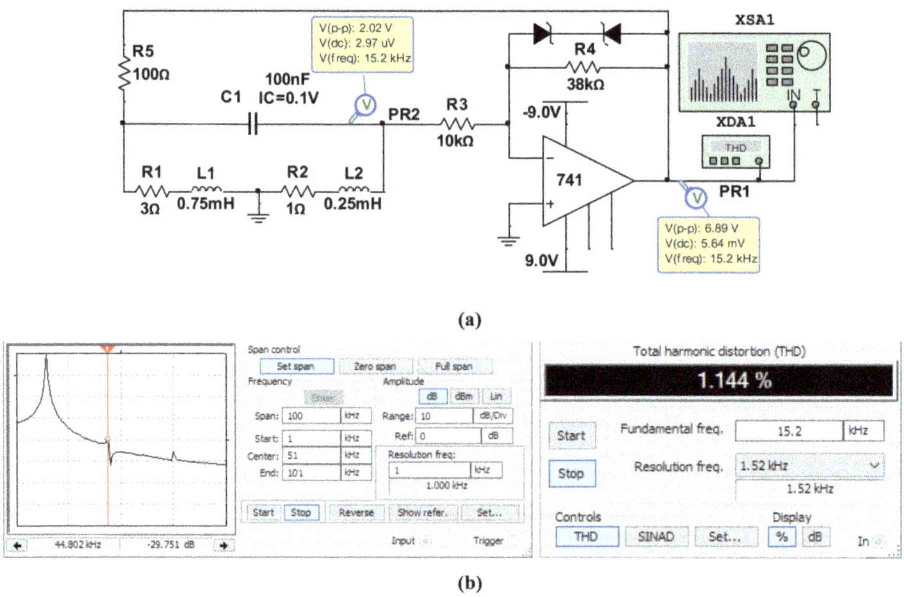

Figure 6.15. LC feedback oscillator with Op-Amp: (a) circuit, (b) measurements.

Figure 6.15(a) illustrates an LC-based oscillator circuit, where the Op-Amp is in an inverting amplifier configuration. In this circuit, a resistor labeled $R5$ is included to aid the startup process and initiate oscillation. However, this resistor is not needed in real-world hardware implementations, where intrinsic noise or other non-idealities naturally trigger startup. During the simulation, a voltage probe can be used to monitor the evolution of the output signal. Initially, the amplitude resides in the picovolt (pV) range, reflecting only background noise. Over time, due to the positive feedback mechanism inherent in oscillator circuits, the amplitude grows exponentially —spanning over twelve orders of magnitude—until it stabilizes in the range of a few volts, limited by the Zener diodes in the circuit. The oscillation frequency is primarily determined by the LC resonant network: $f_o = 1/(2\pi\sqrt{LC}) \approx 15.9\,\text{kHz}$. Minor deviations from the expected frequency are attributed to parasitic capacitances inherent in surrounding components.

Figure 6.15(b) presents the simulation results obtained using a spectrum analyzer and a THD meter. There is a significant improvement from the Wien-bridge oscillator in terms of harmonic content and signal purity. The third harmonic appears at approximately −40 dB relative to the fundamental, demonstrating effective suppression of nonlinear distortion. The simulated THD is approximately 1.14%, which is acceptable for many general-purpose applications. However, if higher spectral purity is required, additional harmonic filtering stages or more linear active elements could be introduced. Overall, these results confirm that the oscillator achieves stable, self-sustained sinusoidal operation with a reasonably low level of harmonic distortion.

> This Op-Amp-based oscillator circuit clearly demonstrates the working principle of this type of oscillator. Unfortunately, the bandwidth of Op-Amps is usually very limited. With the Op-Amp replaced by a transistor, the oscillator can achieve higher-frequency operation and improved power efficiency.

6.4.3 Oscillator based on a transistor

Besides the two pins used for power supply connections, a standard Op-Amp features three functional terminals: one output and two inputs—the inverting and non-inverting inputs. Interestingly, a bipolar junction transistor (BJT) also has three terminals—collector, base, and emitter—which invites a useful analogy between the two devices in terms of signal behavior. If we consider the collector as the output node, the base functions analogously to the inverting input of an Op-Amp, while the emitter resembles the non-inverting input. This comparison becomes particularly evident when analyzing common-emitter (CE) and common-base (CB) amplifier configurations. In the CE amplifier, a signal applied to the base produces an inverted output at the collector, similar to the action of an Op-Amp with a signal at its inverting input. On the other hand, a CB amplifier with an input at the emitter resembles the non-inverting mode in behavior. Although this analogy is not exact in terms of the details, it provides a useful conceptual bridge when transitioning from Op-Amp-based circuits to transistor-based circuits.

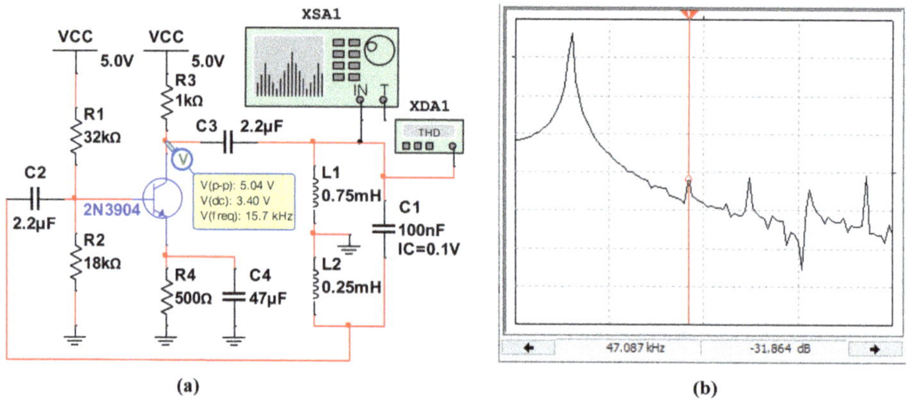

Figure 6.16. LC feedback oscillator with BJT: (a) circuit, (b) spectrum.

Figure 6.16(a) shows a BJT version of the LC feedback oscillator circuit, where the parasitic resistors of the inductors are ignored. Let us concentrate on the LC network; one end is connected to the collector and the other end is connected to the base of the BJT, making it similar to the Op-Amp circuit shown in figure 6.15(a). In addition, the oscillation frequency is still determined by the LC network, and the relatively large coupling capacitors (C_2 and C_3) cause a very limited frequency shift. The spectrum shown in figure 6.16(b) is similar to that of the Op-Amp counterpart shown in figure 6.15(b), and the THD value is also similar.

6.5 Oscillator circuits with LC feedback networks II

The oscillator circuits discussed so far rely on the feedback network to determine the oscillation frequency, while the amplifier itself typically provides broadband gain—that is, the gain remains relatively constant over a wide frequency range. However, this configuration can be inverted: the amplifier itself can be made frequency-selective (tuned amplifier), while the feedback network remains non-selective.

Figure 6.17(a) shows a classic example of a tuned amplifier, where an LC tank circuit is placed in the collector branch of a BJT and serves as the load impedance. In this configuration, the parallel LC circuit exhibits very high impedance at its resonant frequency, ideally approaching infinity if the parasitic resistance is neglected. As a result, the amplifier gain becomes frequency-dependent and exhibits a sharp peak at resonance. This behavior is clearly illustrated in the Bode plot of figure 6.17(b), where the gain response shows a pronounced peak corresponding to the resonant frequency of the LC tank. It can be verified that this peak occurs at the resonant frequency of the LC tank circuit: $f_o \approx 15.9$ kHz.

Building upon the tuned amplifier configuration shown in figure 6.17(a), an oscillator can be formed by introducing a feedback path from the LC tank circuit to the emitter of the BJT, as illustrated in figure 6.18(a). This circuit is commonly referred to as the Hartley oscillator, which is characterized by its use of a split inductor to provide both the frequency-selective load and the necessary feedback. In

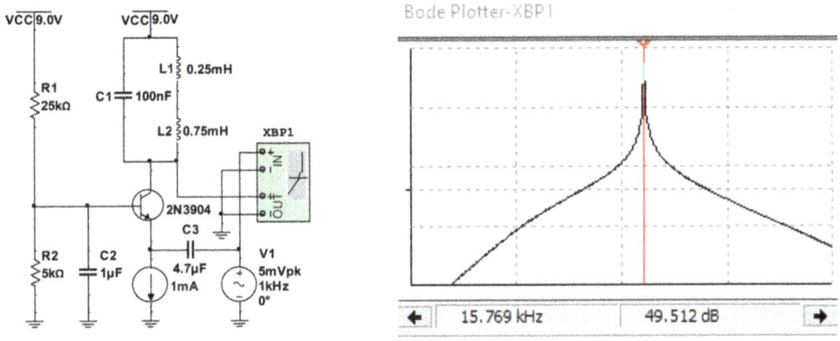

(a) (b)

Figure 6.17. Tuned amplifier: (a) circuit, (b) Bode plot.

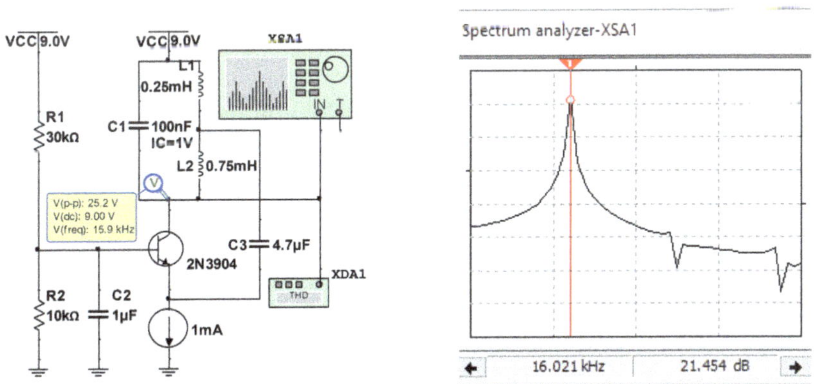

(a) (b)

Figure 6.18. Hartley oscillator implemented using a BJT: (a) circuit, (b) spectrum.

this configuration, the two inductors act as a voltage divider, and the feedback signal is derived from their junction. The emitter is AC coupled to this node, enabling positive feedback that satisfies the Barkhausen criteria for sustained oscillation at the resonant frequency of the tank circuit. The capacitive coupling also ensures that the DC biasing of the transistor remains unaffected.

Figure 6.18(b) shows the frequency spectrum of the oscillator's output. Both the second and third harmonics are significantly attenuated, which indicates high waveform symmetry and effective suppression of nonlinear distortion. The measured THD is approximately 1.2%, confirming the oscillator's ability to generate a clean, low-distortion sine wave. This level of spectral purity is acceptable for many applications.

The Hartley oscillator circuit shown in figure 6.18(a) provides a clear demonstration of the operating mechanism of this topology. However, in practice, a slightly modified version is more commonly used, as shown in figure 6.19(a). In this

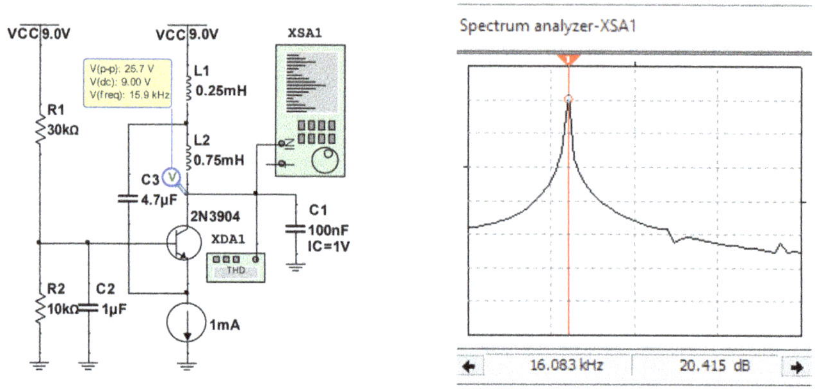

(a) (b)

Figure 6.19. Revised Hartley oscillator based on a BJT: (a) circuit, (b) spectrum.

variation, the configuration is functionally equivalent but optimized for simpler layout and improved grounding.

As is well known from AC circuit analysis, a DC voltage source behaves as an AC ground. This principle allows some flexibility in how the capacitor in the LC tank is connected. In the modified circuit, the 100 nF capacitor C1 is directly connected to ground, simplifying the physical layout and often improving stability in practical implementations.

Despite this structural change, the oscillator's behavior remains virtually unchanged, as demonstrated in figure 6.19(b). The simulated frequency spectrum confirms that the waveform characteristics—such as harmonic content and spectral purity—are preserved. The THD remains at approximately the same level, indicating that the oscillator still produces a high-purity sinusoidal output.

This variation demonstrates the design flexibility of the Hartley oscillator and its resilience to small circuit changes, as long as the fundamental feedback and resonance conditions are satisfied. The configuration in figure 6.19(a) is therefore widely adopted in practical analog and RF oscillator designs where simplicity and compactness are desirable.

> The Hartley oscillator is a classic type of sinusoidal oscillator invented in 1915 by American engineer **Ralph V L Hartley**. Hartley was a pioneer in the field of electronics and information theory, best known for his contributions to early radio communication systems. He introduced this oscillator design while working at the Western Electric Company to generate stable, continuous-wave signals for radio transmission. This oscillator is widely used in RF applications due to its simplicity, ease of tuning, and ability to produce low-distortion sine waves.

As discussed in section 6.3, the LC tank circuit used in oscillators can be configured in two primary ways: split inductor or split capacitor. A few years after the invention of the Hartley oscillator, which employs the split-inductor configuration, Edwin H. Colpitts introduced its complementary design—the Colpitts oscillator—which uses a split-capacitor configuration, as illustrated in figure 6.20(a). In this design, the capacitive voltage divider provides the necessary feedback to sustain oscillation, while the inductor serves as the common reactive element.

The frequency spectrum shown in figure 6.20(b) closely resembles that of the Hartley oscillator, with similarly low levels of harmonic distortion. The measured THD is comparable, confirming that both circuits can generate clean sine waves under appropriate bias and component conditions.

Compared with the Hartley oscillator, the Colpitts oscillator demonstrates superior performance, particularly at higher frequencies. This advantage arises primarily from the favorable high-frequency characteristics of capacitors, which exhibit lower parasitic inductance, higher Q factors, and greater thermal stability than inductors. As a result, Colpitts oscillators offer better frequency stability with lower phase noise, while also being less susceptible to magnetic coupling issues. Moreover, their structure is more conducive to miniaturization, making them especially suitable for RF applications and monolithic integration in modern integrated circuit (IC) and communication circuit designs.

However, one practical drawback of the traditional Colpitts configuration is the complexity of its tuning. Since the frequency of oscillation is determined by the series combination of two capacitors, any frequency adjustment requires changing both capacitors simultaneously while maintaining the correct ratio—a task that is not straightforward in hardware implementations.

To address this limitation, the Clapp oscillator was developed as an enhanced variant of the Colpitts design. In this configuration, an additional capacitor is inserted in series with the inductor in the LC tank. This tuning capacitor becomes the dominant frequency-controlling element, allowing for precise and convenient tuning without disturbing the capacitive feedback ratio. The Clapp oscillator retains the advantages of the Colpitts topology while improving tuning accuracy and

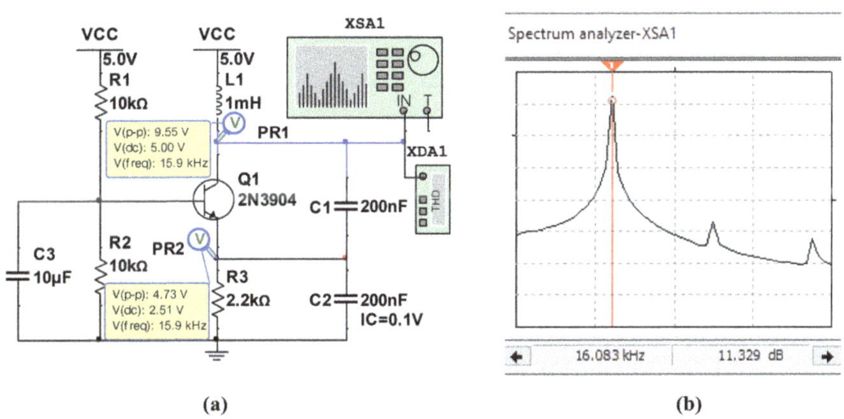

Figure 6.20. Colpitts oscillator based on a BJT: (a) circuit, (b) spectrum.

frequency control, making it well-suited for variable frequency oscillators and frequency synthesizers.

> The Colpitts oscillator is a widely used type of sinusoidal oscillator invented by **Edwin H Colpitts** in 1918. Colpitts was an American engineer and a research director at Western Electric and Bell Telephone Laboratories. He made significant contributions to early radio and telecommunication technologies, and his oscillator design remains fundamental in analog circuit design to this day.

6.6 Oscillator circuits with negative resistance I

As we know, a coupled LC circuit behaves analogously to a mass–spring mechanical system, where oscillation can be initiated when initial energy is supplied to the system. Just as a mass–spring system oscillates due to the transformation between kinetic and potential energies, the inductor and capacitor in an LC circuit exchange energy between magnetic and electric fields. However, resistors in the circuit dissipate electric energy as heat when current flows through them, introducing a damping effect similar to friction in mechanical systems.

Figure 6.21(a) illustrates a parallel RLC circuit; the corresponding voltage signal over time is shown in figure 6.21(b), which is a decaying sinusoidal waveform. This damped oscillation occurs because the resistor continuously reduces the system's energy, causing the amplitude of the waveform to decrease over time. The natural frequency of oscillation is primarily determined by the inductor and the capacitor: $\omega_o = 1/\sqrt{LC}$. Another important parameter is the quality factor Q, which is inversely proportional to the percentage of energy loss per cycle and is related to the parameters of all these devices: $Q = R \cdot \sqrt{C/L}$. Therefore, a higher Q implies lower energy loss and slower oscillation decay, resulting in a waveform that persists longer before damping out. Understanding the relationship between these parameters is fundamental for designing oscillators with the desired stability and spectral purity.

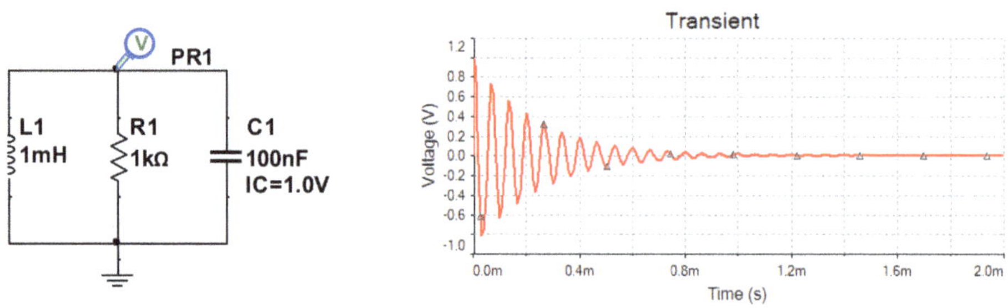

(a) (b)

Figure 6.21. Attenuated oscillation of an RLC circuit: (a) circuit, (b) waveform.

If the resistor in the parallel RLC circuit shown in figure 6.21(a) is removed, the system becomes an ideal LC circuit with no explicit damping. In theory, this would result in perpetual oscillation, as energy would continuously transfer between the inductor and capacitor without loss. However, this scenario is not physically realistic. In practice, every inductor inherently possesses some resistance due to the finite conductivity of the wire used in its windings. These windings, typically made of thin copper wire wrapped around a magnetic core, introduce a small but non-negligible series resistance, which acts as an internal damping element and gradually dissipates energy.

To better understand the practical behavior of the system, we must account for this parasitic resistance. The original circuit in figure 6.22(a), which includes such realistic components, is relatively complex and not easily amenable to direct analysis. Therefore, it is often transformed into an equivalent circuit, as shown in figure 6.22(b), which simplifies the analysis while preserving the essential electrical characteristics.

The fundamental requirement for this transformation is that both circuits exhibit the same impedance at the resonant frequency of interest: $Z_S = Z_P$. In addition, the quality factor should also remain unchanged: $Q = \omega_0 L_s / R_s = \omega_0 C R_p$. Using these conditions, the parameters of the equivalent parallel circuit can be derived from the original series circuit:

$$L_p = \left(1 + \frac{1}{Q^2}\right)L_s, \quad R_p = \left(1 + Q^2\right)R_s. \tag{6.10}$$

For circuits with a high quality factor, $Q > 5$, the two transformation equations can be simplified considerably: $L_p \approx L_s$, $R_p \approx Q^2 R_s = L_s/(R_s C)$. These results highlight that the equivalent parallel resistance is inversely proportional to the original series resistance. This transformation is particularly useful in designing and analyzing oscillator feedback networks, where high-impedance parallel resonant circuits are preferred for their sharper frequency selectivity and better compatibility with voltage-mode amplifiers.

Although the parasitic resistance of an inductor cannot be entirely eliminated, its damping effect can be counteracted by introducing a negative resistance into the circuit. While conventional (positive) resistance dissipates electrical energy as heat,

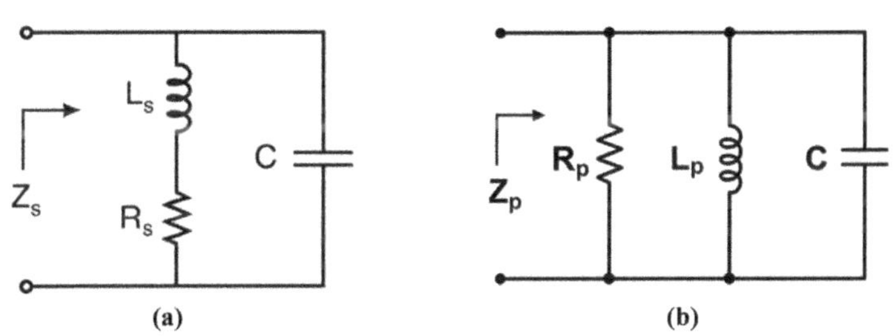

Figure 6.22. Transformation of RLC circuits: (a) series resistance, (b) parallel resistance. Created with GPT-4.0, OpenAI.

negative resistance, by contrast, supplies energy to the circuit, thereby sustaining oscillation. This behavior cannot be achieved with passive components alone; instead, active devices must be employed to synthesize negative resistance. This principle forms the basis for negative-resistance oscillators, which are widely used in RF and microwave applications due to their ability to produce stable oscillations without relying on traditional feedback configurations.

Figure 6.23(a) illustrates an equivalent circuit in which an Op-Amp is configured to generate negative resistance. A test voltage source is applied to the input, and a current probe measures the response. The direction of the measured current—flowing opposite to what would be expected for a positive resistor—indicates that the circuit behaves as if it has negative resistance. Applying Thévenin's theorem allows us to obtain the input impedance seen by the test source: $R_{\text{in}} = -5\ \text{k}\Omega$. Using circuit analysis, the formula for the input resistance can be derived:

$$R_{\text{i}} = V_{\text{i}}/I_{\text{i}} = -\frac{R_1}{R_2}R_3. \tag{6.11}$$

This behavior is further illustrated in figure 6.23(b), which plots the I–V characteristics at the input node. The negative slope of the curve confirms the presence of negative resistance. Such a slope implies that an increase in voltage leads to a decrease in current, a hallmark of negative resistance. Circuits exhibiting this property are essential in the design of active resonators and oscillators, where they effectively cancel out losses and enable sustained sinusoidal oscillation.

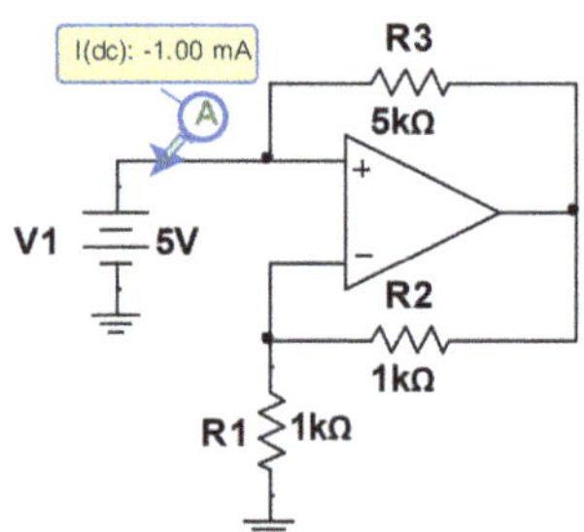

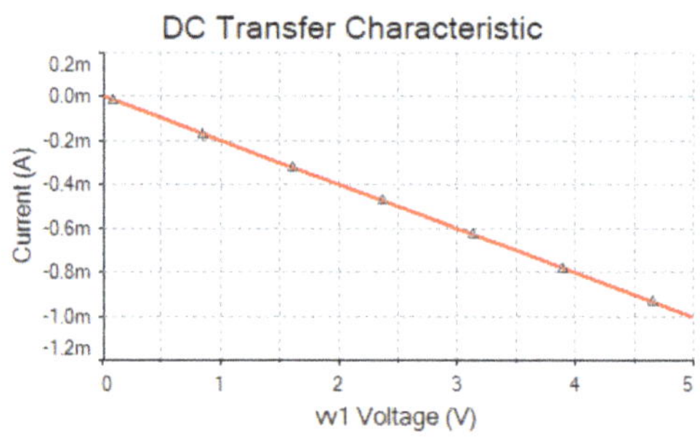

Figure 6.23. Op-Amp circuit with negative resistance: (a) circuit, (b) I–V characteristics.

Figure 6.24(a) illustrates an oscillator circuit that incorporates a negative resistance to compensate for energy losses caused by a parasitic resistance in the inductor. Specifically, the 10 kΩ parallel resistor in the equivalent model arises from the transformation of a 1 Ω series parasitic resistance associated with the inductor. To sustain oscillation, the total resistance of the resonant circuit must be negative, or at least zero, to counteract the energy dissipation from the parasitic resistance. This is achieved by introducing a negative resistance in parallel with the equivalent positive resistance.

To ensure that the net parallel resistance becomes negative—with a net energy gain—the magnitude of the negative resistance must be **smaller** than the value of the

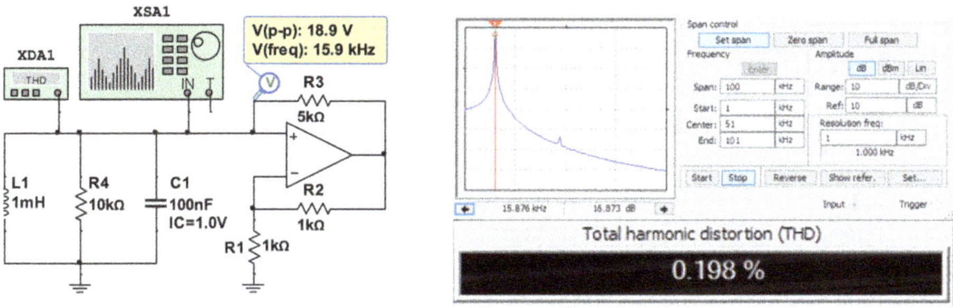

(a) (b)

Figure 6.24. Oscillator with negative resistance: (a) circuit, (b) spectrum and THD.

equivalent positive resistance. This condition can be understood from the formula for the total resistance of two resistors in parallel: $1/R_T = 1/R_p + 1/R_n$. In this circuit, $R_p = 10\ \text{k}\Omega$, $R_n = -5\ \text{k}\Omega$, and $R_T = -10\ \text{k}\Omega$.

Figure 6.24(b) presents the simulated output spectrum and THD of this negative-resistance oscillator circuit. Notably, the spectrum exhibits a clean sinusoidal waveform with a dominant fundamental frequency component. The second harmonic is effectively suppressed, as evidenced by the absence of a visible peak at twice the fundamental frequency. Moreover, the third harmonic, though present, is extremely small—measured at approximately 50 dB below the fundamental peak. This indicates a high degree of waveform purity.

The calculated THD is approximately 0.2%, which is significantly lower than that of the oscillator circuits analyzed earlier. This low-distortion performance is a key advantage of negative-resistance oscillator designs, which can produce highly stable and spectrally pure signals due to their linear operating characteristics and minimal nonlinearity in the active components. Such characteristics make negative-resistance oscillators particularly suitable for high-frequency and precision applications where signal integrity and low harmonic content are critical.

6.7 Oscillator circuits with negative resistance II

Although the oscillator circuit discussed in the previous section demonstrates excellent spectral purity and low distortion, its reliance on an Op-Amp imposes certain limitations. Specifically, Op-Amps typically have limited bandwidth and slew rate, which restrict their usefulness in high-frequency applications. Additionally, they tend to consume more power and occupy more chip area in integrated designs. Therefore, for higher-frequency or more power-sensitive applications, a simpler and more efficient alternative that can still exhibit negative resistance is desirable.

Figure 6.25(a) presents a cross-coupled BJT circuit, a configuration widely used for implementing negative resistance without requiring an Op-Amp. This topology shares structural similarity with a differential amplifier, but it operates in a different way. In this setup, the voltage source on the left acts as a test input that probes the impedance characteristics of the circuit, while the voltage source on the right—with a 180° phase shift—is included for symmetry and balance, ensuring equal biasing and consistent operation of both transistors.

Figure 6.25(b) shows the voltage and current waveforms at the input node, which exhibit a 180° phase shift—a clear indicator of negative resistance. This antiphase relationship means the current flows in the opposite direction to that in a passive element, confirming that the circuit supplies energy rather than dissipating it, thus enabling sustained oscillation. Similar behavior appears symmetrically on the right-hand branch and can also be observed between the branches when a test voltage source is inserted.

In a similar manner, oscillation arises when an LC tank circuit is connected to the negative-resistance core. As shown in figure 6.26(a), the LC tank is inserted between

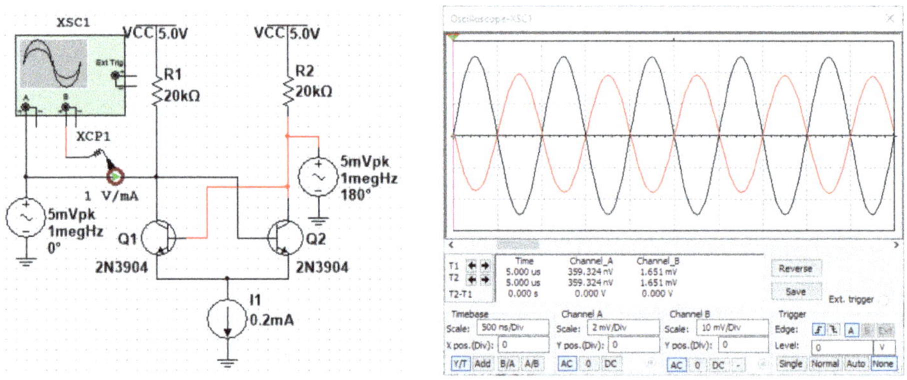

Figure 6.25. Negative resistance produced by a cross-coupled BJT circuit: (a) circuit, (b) waveform.

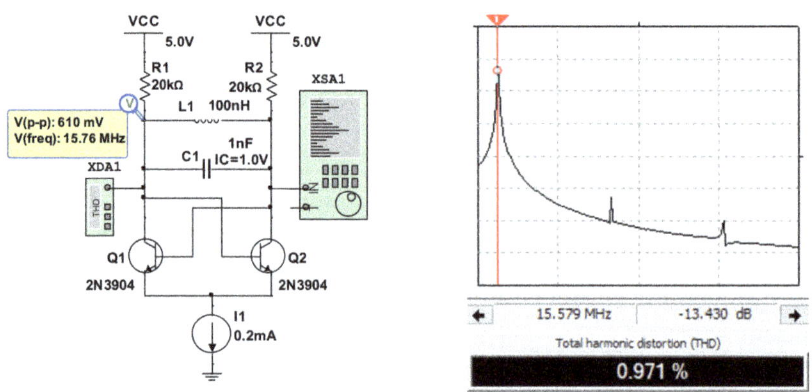

Figure 6.26. Oscillator with cross-coupled negative resistance: (a) circuit, (b) spectrum and THD.

the two branches of the cross-coupled circuit. This configuration enables energy exchange between the tank and the active circuit, sustaining oscillation. However, the voltage probe measurements reveal that the actual oscillation frequency slightly deviates from the ideal LC resonant frequency, primarily due to parasitic capacitances associated with the BJTs.

Figure 6.26(b) displays the frequency spectrum and THD of the output signal. There are no even-order harmonics, and the third and fifth harmonic components are also very weak. The THD measures approximately 0.97%, indicating a low level of nonlinearity. The overall performance is very good.

As is well known, the presence of parallel resistance in an RLC tank circuit lowers the quality factor (Q) and degrades oscillation purity by increasing the energy loss per cycle. In the circuit shown in figure 6.26(a), the two 20 kΩ resistors located at the top effectively act as such parallel resistances, thus diminishing the tank's Q-factor. To enhance performance, these resistors can be eliminated.

Figure 6.27(a) presents an improved version of the circuit with the resistors removed. Additionally, the split inductor in the tank circuit resembles the topology discussed earlier in section 6.3, where magnetic coupling and symmetry improved overall efficiency. Voltage probe measurements reveal that the amplitude and frequency remain comparable to the previous design, confirming that the removal of the resistors does not negatively impact the oscillation. Figure 6.27(b) illustrates the resulting frequency spectrum and THD of the improved circuit. First, the spectrum looks similar to the previous one shown in figure 6.26(b). Second, the THD is reduced from 0.97% to 0.71%, reflecting lower nonlinear distortion.

Due to their compact structure, tunability, and ability to operate efficiently at high frequencies, cross-coupled oscillators are widely used across a broad range of analog and RF systems. Their ability to generate clean and stable oscillations makes them especially valuable in ICs, where space, power efficiency, and performance are critical.

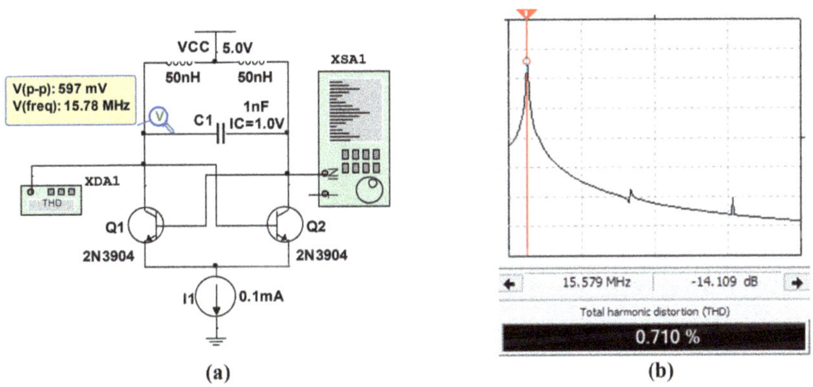

Figure 6.27. Oscillator with cross-coupled negative resistance: (a) circuit, (b) spectrum and THD.

Negative resistance can also be achieved using devices such as the Gunn diode, which relies on the transferred electron effect in materials such as GaAs. When biased into its negative-resistance region, the Gunn diode can generate microwave and millimeter-wave oscillations, often reaching frequencies above 100 GHz. Gunn diode oscillators are widely used in radar, wireless communication, and microwave test systems due to their simplicity, solid-state reliability, and ability to operate at very high frequencies.

6.8 Coupled oscillator circuits

When two oscillators are coupled together, they can exhibit two fundamental modes of oscillation: symmetric or antisymmetric. This behavior is analogous to that of a coupled mass–spring system, as illustrated in figure 6.28. If the coupling spring at the center is removed, each mass (open circle) is attached to the wall with a spring. Focusing on the two masses, the system can oscillate in two distinct ways. In the symmetric mode, shown in figure 6.28(a), the two masses move in the same direction. In contrast, the antisymmetric mode, illustrated in figure 6.28(b), involves both masses moving in opposite directions simultaneously. These two modes correspond to the normal modes of the system.

A particularly interesting feature of the symmetric mode is that the coupling spring—the one connecting the two oscillators—is neither stretched nor compressed during the motion, assuming the two oscillators are identical. As a result, the coupling element does not contribute to the restoring force in this mode, and the oscillators behave as if they were independent but synchronized. Therefore, the oscillation frequency remains unchanged: $\omega_o = \sqrt{k/m}$. In contrast, the antisymmetric mode involves maximum engagement of the coupling spring, and the oscillation frequency becomes $\omega_o = \sqrt{(k + 2k_c)/m}$, where k_c stands for the stiffness of the coupling spring and the factor of two is due to the stretching or compression on both sides. In other words, the deformation of this coupled spring is twice as much as that of the spring on either side. These principles directly translate to coupled electrical oscillators, where the behavior of coupling elements such as capacitors or inductors similarly distinguishes the symmetric and antisymmetric modes of oscillation.

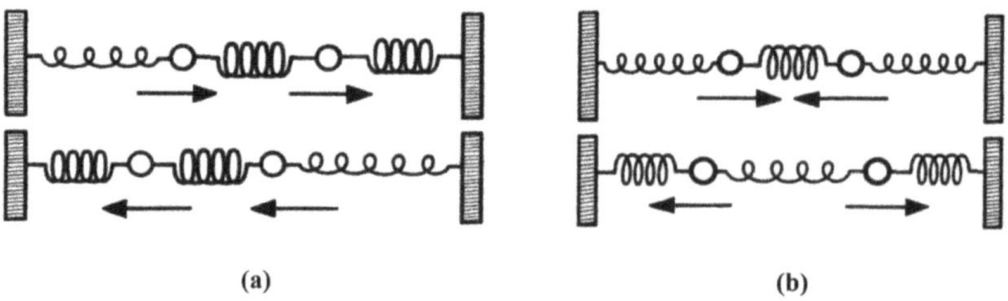

Figure 6.28. Coupled oscillators: (a) symmetric mode, (b) antisymmetric mode. Created with GPT-4.0, OpenAI.

6.8.1 Symmetric oscillation mode

Figure 6.29(a) illustrates a circuit composed of two identical oscillators coupled via a capacitor, with initial conditions configured to excite the symmetric mode of oscillation. Unlike the leftward and rightward motions of the mass–spring system, here, the voltage goes up and down. The polarities of the two capacitors (C_1 and C_2) are opposite, as marked by the plus signs (+) next to the capacitors. From the perspective of the coupling capacitor C_3, this results in zero voltage across its terminals.

Because the two oscillators are identical and oscillate in synchrony, this condition persists throughout the oscillation. In effect, the coupling element becomes electrically inactive, and the interaction between the two oscillators appears negligible under symmetric-mode excitation. In this case, the oscillation frequency is not affected by the coupling, as indicated in the simulation results shown in figure 6.29(a).

Figure 6.29(b) displays four waveforms corresponding to the signals measured at the four oscilloscope probe points, ordered from left to right. Focusing on the two middle traces—representing the voltages at both ends of the coupling capacitor—we observe that the signals are in phase and exhibit equal amplitude. This confirms the presence of symmetric-mode oscillation, as previously analyzed. Furthermore, within each oscillator, the two branches (traces 1 and 2 for one oscillator, and traces 3 and 4 for the other) show signals that are 180° out of phase but equal in amplitude. This feature is characteristic of balanced oscillator operation and indicates that the oscillators remain unaffected by the coupling capacitor, consistent with the fact that no net current flows through it during symmetric-mode oscillation.

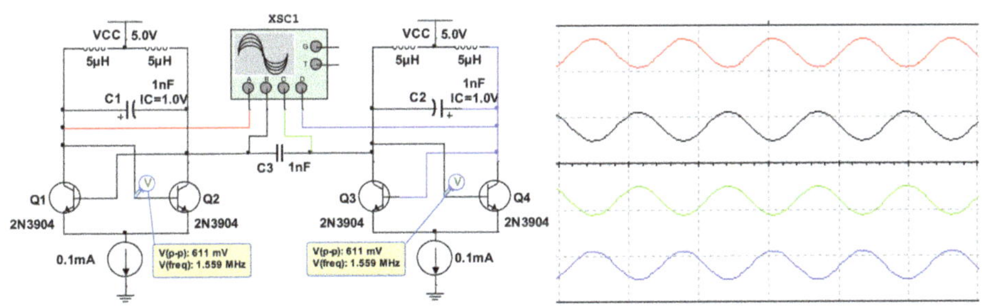

(a) (b)

Figure 6.29. Oscillators coupled with a capacitor in symmetric mode: (a) circuit, (b) waveforms.

Figure 6.30 presents a circuit configuration nearly identical to that in figure 6.29, with the only difference being the coupling element: the capacitor is replaced by an inductor. However, under symmetric-mode oscillation, there is essentially no net coupling between the oscillators, regardless of whether a capacitor or an inductor is used. This is because, in the symmetric mode, the voltages at both ends of the coupling element remain equal, resulting in no voltage difference across the coupling element. Consequently, the characteristics of the coupling element have negligible influence on the circuit behavior in this mode.

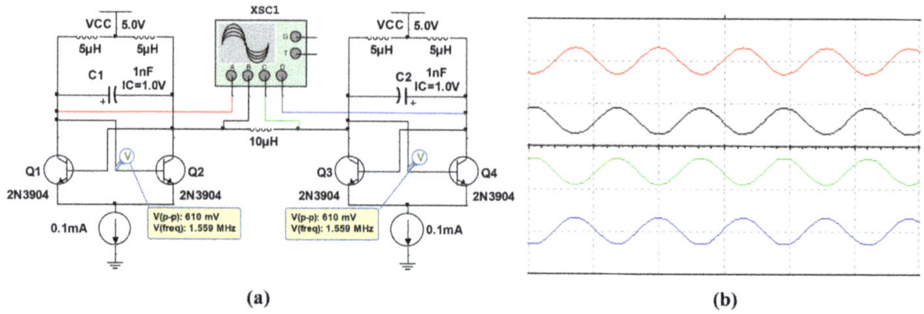

Figure 6.30. Oscillators coupled by a capacitor in symmetric mode: (a) circuit, (b) waveforms.

As a result, the waveforms observed in figure 6.30(b) are nearly identical to those shown in figure 6.29(b), further confirming that the two oscillators operate independently and in phase. Additionally, the oscillation frequency and amplitude remain unchanged, as illustrated in figure 6.30(a). This insensitivity to the coupling device in symmetric-mode oscillation underscores a key feature of coupled systems: the effect of the coupling element strongly depends on the mode of excitation.

6.8.2 Antisymmetric oscillation mode

To excite the antisymmetric mode of oscillation, it is sufficient to modify the initial conditions of the two capacitors, as illustrated in figure 6.31(a). By assigning equal initial voltages to C_1 and C_2, differential excitation is established across the coupling inductor, which drives the system into the antisymmetric mode. The resulting waveforms are shown in figure 6.31(b), with the traces corresponding to the four oscilloscope probes ordered from left to right.

Examining the two middle traces, which represent the voltages at both ends of the coupling inductor, we observe that the signals have equal amplitude but are 180° out of phase. This clear phase opposition is a defining characteristic of antisymmetric oscillation and indicates that the coupling inductor is now actively involved between the two oscillators. Unlike the symmetric mode—where the coupling element remains electrically inactive—in the antisymmetric mode, the inductor facilitates strong interaction, storing and releasing energy cyclically between the two oscillating nodes.

The antisymmetric waveforms observed in figure 6.31(b) imply that the coupling inductor functions analogously to a balanced seesaw, where the voltages at each end swing in opposite directions. This behavior allows us to conceptually divide the 10 μH coupling inductor into two equal 5 μH inductors, with the central node grounded in AC. In this circuit, it is connected to a 5 V DC voltage source, which matches V_{CC} so that no DC current flows.

With this transformation, the system can be analyzed using an equivalent half-circuit model, effectively decoupling the original configuration into two symmetric subcircuits. Figure 6.32(a) shows the half-circuit on the left side, where the simulation results confirm the validity of this transformation. In the context of

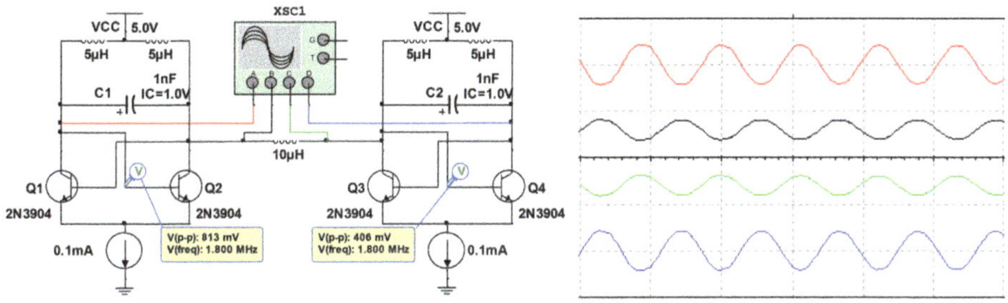

(a) (b)

Figure 6.31. Inductively coupled oscillators in antisymmetric mode: (a) circuit, (b) waveforms.

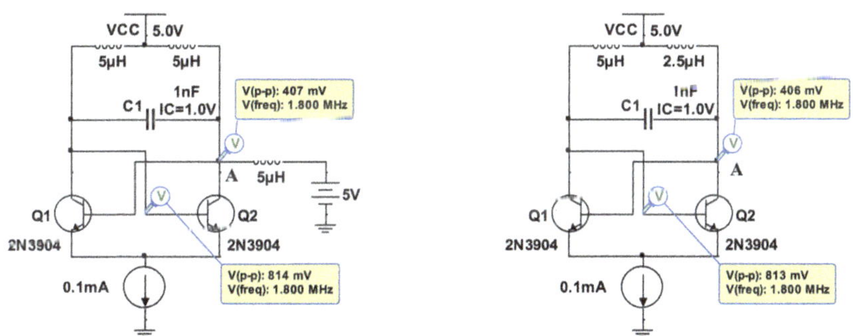

(a) (b)

Figure 6.32. Equivalent oscillator circuits: (a) circuit with a half-coupling inductor, (b) transformed circuit.

AC analysis, the DC supply voltage (V_{CC}) behaves as an AC ground, enabling further simplification of this equivalent circuit.

From the perspective of node *A* on the right branch, it is connected to two AC-grounded 5 μH inductors. Since these inductors are in parallel, their combined inductance is reduced to 2.5 μH, as shown in figure 6.32(b). The total inductance in parallel with the capacitor in the LC tank now becomes 7.5 μH. Paired with the capacitor in the tank circuit, this yields a theoretical resonant frequency of approximately 1.84 MHz, calculated using the standard LC resonance formula. Simulation results from the voltage probes indicate an actual oscillation frequency of 1.80 MHz, which closely matches the theoretical value.

As discussed in section 6.3, when the inductor in an LC tank is unevenly divided, the voltage amplitudes at the ends of the split inductor scale in proportion to their respective inductance values. This principle applies directly to the circuit configuration shown in figure 6.32(b), where the inductance is asymmetrically distributed between the two branches. Specifically, the inductance is 5 μH on the left and 2.5 μH on the right.

In the simulation, the voltage probe at the bottom captures the signal on the left side of the LC tank and records a peak-to-peak amplitude of 813 mV. In contrast, the voltage probe at the top measures the signal on the right side and shows a peak-to-peak amplitude of 406 mV. The measured amplitude ratio of approximately 2:1 aligns closely with the 2:1 inductance ratio between the left and right segments.

If the coupling inductor is replaced with a coupling capacitor, the behavior of the coupled oscillator circuit becomes more interesting—but also more complex. Figure 6.33(a) shows such a configuration, where two oscillators are coupled through a 1 nF capacitor, and the resulting waveforms are displayed in figure 6.33(b), corresponding to the four voltage probes arranged from left to right on the oscilloscope.

As discussed for the inductively coupled oscillator circuit, the initial conditions of capacitors C_1 and C_2 are the same. From the perspective of the coupling capacitor C_3, this represents antisymmetric excitation. Such an initial condition naturally favors the differential mode of oscillation, leading to a 180° phase shift between the voltages on both sides of C_3. This antiphase relationship is clearly observed in the two traces located in the middle of the waveform diagram shown in figure 6.33(b), where the waveforms are mirror images of each other.

Following the same analytical approach used earlier, the 1 nF coupling capacitor can be thought of as two 2 nF capacitors in series, with their central node connected to ground. This transformation allows the circuit to be conceptually decoupled so that each oscillator can be analyzed independently, and each sees a grounded 2 nF capacitor connected to one branch of the circuit.

Examining the first two traces in figure 6.33(b), we again observe a 180° phase shift between the waveforms, indicating that the two nodes swing in opposite directions, much like the motion of a seesaw. This phase relationship suggests that capacitor C_1 can be split into two capacitors in series, labeled C_{out} (outside) and C_{in} (inside), forming an uneven capacitive divider, as shown in figure 6.34(a). This is necessary because the waveform amplitudes in the two branches are unequal, implying asymmetric capacitive loading. Furthermore, since both C_{in} and C_3 are connected to the right branch of this circuit, they effectively form a parallel combination. As a result, the total equivalent capacitance on the right side becomes $C_R = C_{in} + 2\,\text{nF}$, which is shown in figure 6.34(b).

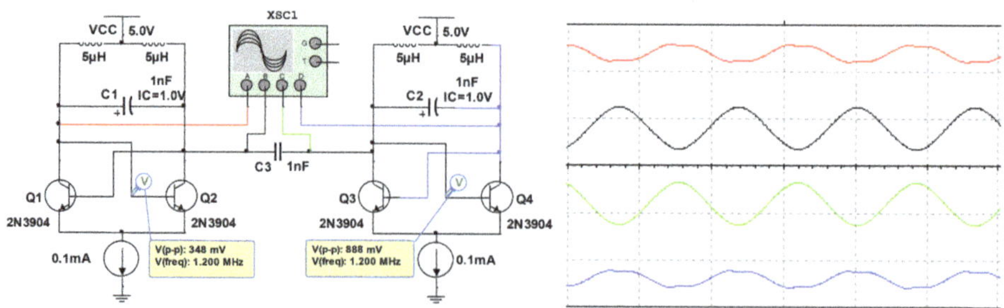

(a) (b)

Figure 6.33. Capacitively coupled oscillators in antisymmetric mode: (a) circuit, (b) waveforms.

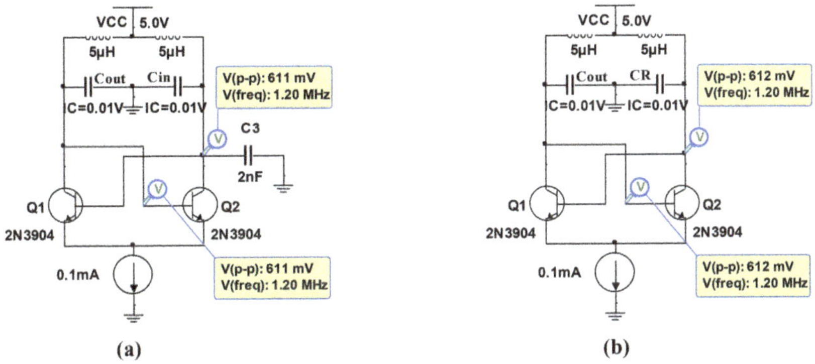

Figure 6.34. Equivalent circuits with a partitioned capacitor.

In the circuit shown in figure 6.34(b), the overall LC tank can be conceptually divided into two coupled LC tanks, since the midpoints of both the inductor pair and the capacitor pair are connected to ground. Under the constraint that both tanks must oscillate at the same frequency, the partitioning scheme used for capacitor C_1 must satisfy this condition, which is listed as the second equation below:

$$\begin{aligned} \frac{1}{C_{\text{out}}} + \frac{1}{C_{\text{in}}} &= \frac{1}{C_1} \\ \omega_o = \frac{1}{\sqrt{LC_{\text{out}}}} &= \frac{1}{\sqrt{L(C_{\text{in}} + 2nF)}}. \end{aligned} \tag{6.12}$$

The second equation provides a simple relationship: $C_{\text{out}} = C_{\text{in}} + 2\text{nF}$. Combined with the first equation, the solution can be found: $C_{\text{in}} = \sqrt{2}$ nF. Substituting this result back into the resonant frequency equation yields a calculated oscillation frequency of 1.22 MHz. The simulated oscillation frequency is 1.20 MHz, as shown in figures 6.33(a) and 6.33(b), which is in close agreement with the calculated value. This excellent match between simulation and theory confirms not only the correctness of the capacitor partitioning approach but also validates the coupled oscillator model and the accuracy of the equivalent circuit analysis.

A closer examination of the waveforms in figure 6.33(b) reveals an interesting and somewhat counterintuitive phenomenon: the inside branches of the circuit exhibit larger amplitude waveforms with significantly lower distortion, whereas the outside branches display smaller amplitude waveforms that are visibly clipped at both the top and bottom. This observation suggests that the inner nodes benefit from stronger coupling, which helps maintain waveform linearity and suppress harmonic distortion.

This behavior implies that enhancing the coupling between oscillators can effectively reduce distortion across the circuit. To confirm this, an additional coupling capacitor is introduced between the two outer branches, as illustrated in figure 6.35(a). This modification strengthens the interaction between these nodes, balances the load distribution, and improves signal symmetry. As a result, the

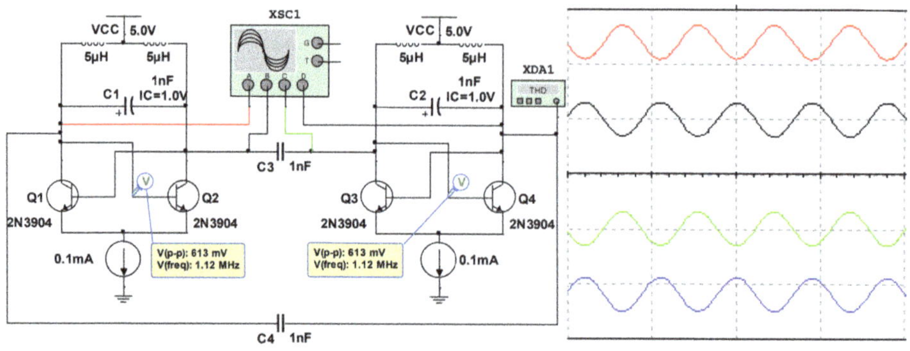

(a) (b)

Figure 6.35. Capacitively coupled oscillators in antisymmetric mode: (a) circuit, (b) waveforms.

waveforms shown in figure 6.35(b) demonstrate noticeably improved linearity, with a simulated THD reduced to only 1.2%, confirming the effectiveness of this coupling strategy in mitigating waveform distortion and enhancing overall oscillator performance.

Following the same approach used to analyze the circuit in 6.30(a), each 1 nF coupling capacitor (C_3 or C_4) can be conceptually split into two 2 nF capacitors in series, with their junction tied to ground. Focusing on one oscillator unit, each branch now contains a grounded 2 nF capacitor, which electrically behaves as two capacitors in series between the two branches, since the ground is a common node. These two grounded capacitors effectively combine into a single 1 nF capacitor, which is in parallel with the local tank capacitor (C_1 or C_2) within the oscillator, increasing the total capacitance in each LC tank to 2 nF.

With this revised capacitance value, the resonant frequency of the oscillator circuit can be calculated. For values of $L = 10\ \mu\text{H}$ and $C = 2$ nF, the calculated resonant frequency is approximately 1.11 MHz, which aligns closely with the simulated result of 1.12 MHz shown in figure 6.35(a). This strong agreement between analytical prediction and simulation further validates the accuracy of the circuit model and the effectiveness of the modified coupling network in maintaining predictable oscillation behavior.

6.8.3 Competition between oscillation modes

In practical experiments, the initial conditions of the two capacitors cannot be explicitly specified or controlled. This raises the question of which oscillation mode naturally dominates. Observations indicate that the symmetric (in-phase) mode is generally preferred, as illustrated in figure 6.36(a). In this setup, the component parameters are $L = 1$ mH and $C = 220$ nF, with a single coupling capacitor connected between the two oscillators.

The preference for the symmetric mode arises from the fact that, in this mode, the voltage difference across the coupling capacitor is ideally zero. As a result, the coupling device carries minimal current, which leads to lower reactive loading and

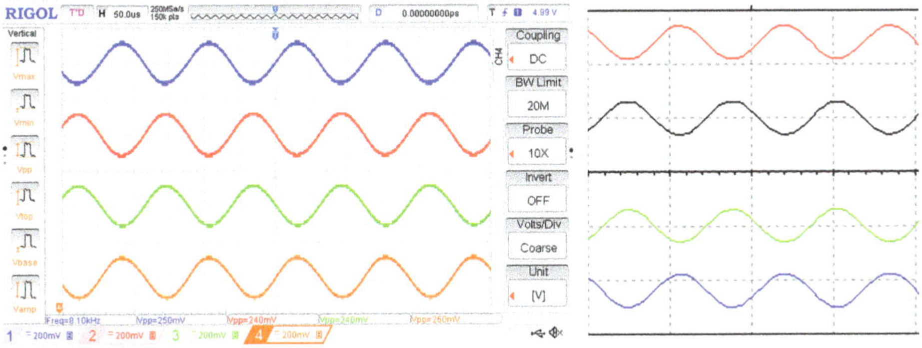

(a) (b)

Figure 6.36. Capacitvely coupled oscillators: (a) experimental results, (b) simulation results.

reduced energy loss in the coupling path. This creates more favorable conditions for sustaining oscillation compared to the antisymmetric (out-of-phase) mode, where the coupling capacitor experiences maximum voltage swing and thus higher reactive current and loading.

Simulation results further confirm this behavior. When the initial voltages across the two capacitors are neither perfectly symmetric nor antisymmetric, the circuit still naturally evolves toward the symmetric (in-phase) oscillation mode, as shown in figure 6.36(b). This outcome demonstrates that the symmetric mode is the more energetically favorable state under typical startup conditions. This preference arises because the symmetric mode minimizes reactive energy exchange between the coupled elements, leading to lower-loss, more stable oscillation.

An instructive analogy is the two possible states of electrons in a hydrogen molecule (H_2), where the bonding state, characterized by constructive wavefunction interference, is energetically favored over the antibonding state, which results from destructive interference. Just as electrons tend to occupy the bonding orbital to achieve a lower energy configuration, the coupled oscillator system similarly gravitates toward the symmetric mode as the natural, stable solution. This analogy highlights a fundamental physical principle observed across disciplines: systems coupled through energy-sharing mechanisms often self-organize into configurations that minimize total system energy.

6.8.4 Coupling of nonidentical oscillators

When the two oscillators have slightly different natural oscillation frequencies, the behavior of the system becomes more complex. Under conditions of weak coupling, the oscillators are unable to synchronize to a single frequency. Instead, the waveform exhibits a characteristic beat-frequency phenomenon, which arises from the superposition of two sine waves oscillating at slightly different frequencies. This results in a periodic modulation of the signal amplitude at a frequency equal to the difference between the two natural frequencies: $f_{\text{beat}} = |f_1 - f_2|$.

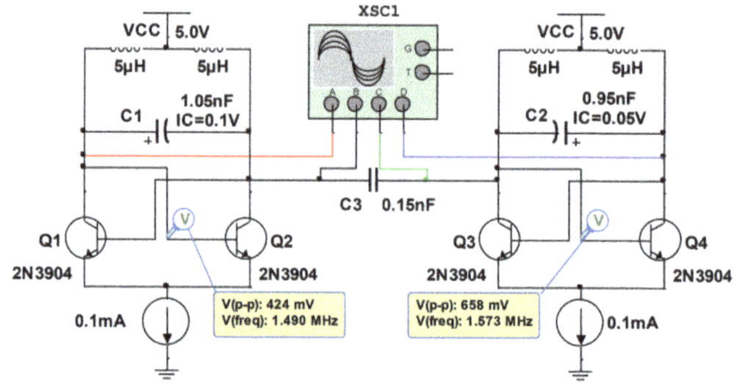

(a)

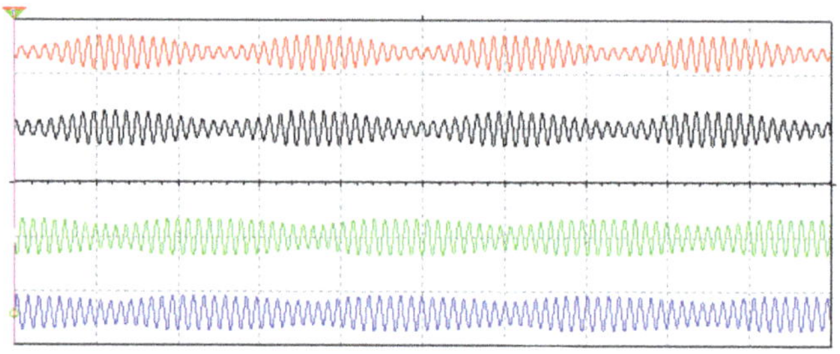

(b)

Figure 6.37. Capacitively coupled nonidentical oscillators: (a) circuit, (b) waveform.

An example circuit configuration is shown in figure 6.37(a), where the capacitances are intentionally mismatched: $C_1 = 1.05$ nF, $C_2 = 0.95$ nF, representing a 5% deviation from the nominal value. The natural resonant frequencies are 1.55 and 1.63 MHz, respectively. Additionally, the coupling capacitance is reduced to $C_3 = 0.15$ nF to ensure weak coupling. Under these conditions, the simulation results in figure 6.37(a) indicate that the two oscillators exhibit distinct natural oscillation frequencies of approximately 1.490 and 1.573 MHz, respectively.

The simulated waveforms, shown in figure 6.37(b), clearly demonstrate the beat phenomenon. The signal oscillates at a frequency near the average of the two natural frequencies, but its amplitude rises and falls periodically at the beat frequency of approximately 83 kHz (the difference between 1.573 and 1.490 MHz). This beat pattern confirms the expected behavior of weakly coupled oscillators with slight frequency detuning, where energy is periodically exchanged between the two oscillators rather than maintaining a stable phase-locked condition.

As the coupling strength increases, the interaction between the two oscillators becomes stronger, leading to frequency locking, where both oscillators synchronize to a common frequency despite their natural frequency difference. This transition from independent oscillation to synchronization is governed by the interplay between the natural frequency detuning and the strength of the coupling.

The circuit shown in figure 6.38(a) is essentially the same as the one in figure 6.37 (a), with the only modification being that the coupling capacitance is increased from

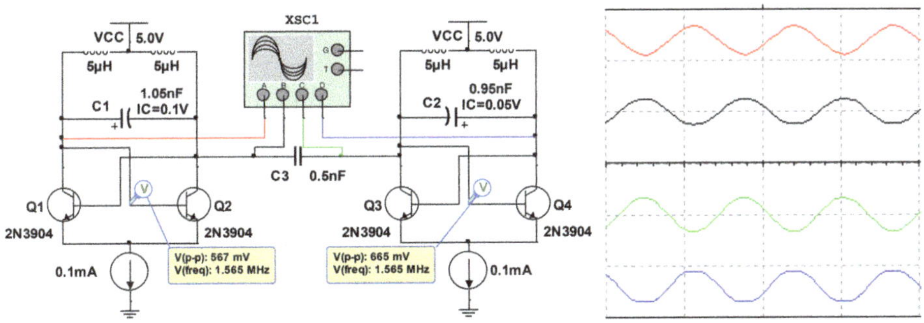

(a) (b)

Figure 6.38. Frequency locking in capacitively coupled oscillators: (a) circuit, (b) waveform.

0.15 to 0.5 nF. With this significantly stronger coupling, the two oscillators are able to overcome their initial frequency mismatch and lock into a single common frequency. The original frequency difference—caused by the 5% mismatch in C_1 (1.05 nF) and C_2 (0.95 nF)—is effectively eliminated.

The simulated waveforms, shown in figure 6.38(b), clearly confirm this behavior. Both oscillators now oscillate at the same frequency, with no observable beat phenomenon. Furthermore, although the initial voltages across capacitors C_1 and C_2 are asymmetric, the system naturally evolves into the symmetric (in-phase) mode. This again highlights the inherent tendency of capacitive coupling to favor symmetric oscillation, especially when the coupling strength is sufficiently high to enforce frequency locking.

Coupled oscillators have broad applications in engineering and science. They are fundamental in RF filters, frequency synthesizers, and PLLs for communication systems. In sensor networks, coupled oscillators enable synchronization and signal coherence. They also play key roles in microwave resonator arrays, clock distribution in ICs, and even in modeling natural phenomena such as biological rhythms, neural networks, and mechanical synchronization in structures and devices.

6.9 Multivibrator circuits

In nature and human society, oscillation phenomena are ubiquitous, but the clean sine wave discussed so far is quite rare. For example, the rhythmic beating of the human heart, the firing patterns of neurons, the rise and fall of predator-prey populations, and the cycles of economic growth and recession are all examples of oscillatory behavior. However, these natural oscillations are often nonlinear, irregular, or involve complex interactions with the environment. Unlike the idealized sine wave produced by a linear LC oscillator, real-world oscillations frequently exhibit waveform distortions, amplitude fluctuations, or chaotic behavior due to damping, external perturbations, or nonlinear feedback. The diversity of oscillatory phenomena reflects the underlying principles of energy exchange and feedback, which manifest broadly in both physical systems and complex networks in biology, sociology, and economics.

The term ‘multivibrator’ refers to a class of oscillator circuits that generate non-sinusoidal waveforms, typically square waves, which are widely used in timing, pulse generation, and switching applications. There are three types of multivibrators: astable, monostable, and bistable. An astable multivibrator has no stable state and continuously oscillates between two states without any external trigger, functioning as a free-running oscillator. Relaxation oscillators belong to this category; they rely on the periodic charging and discharging of a capacitor to produce a square-wave output, as illustrated in figure 6.39.

At the top of the relaxation oscillator circuit, resistors R_1 and R_2 form a positive feedback network that forces the output to switch between the positive and negative supply rails. In this circuit, an ideal Op-Amp is used, so the output voltage can reach the rail voltages at ±5 V. In addition to providing positive feedback, R_1 and R_2 also define the upper and lower threshold voltages that control when the output switches from one state to the other. In this circuit, R_1 and R_2 are equal, so the respective thresholds are half of the output rail voltages, i.e. ±2.5 V. This is similar to the Schmitt trigger circuit discussed in chapter 5.

The capacitor connected to the inverting input charges and discharges through a separate resistor, typically controlling the oscillation frequency. When the capacitor voltage crosses the threshold, the output abruptly switches polarity, creating the characteristic square-wave output. Due to the symmetry between the positive and negative swings, the duty cycle of the square wave is 50%, meaning the high and low phases are equal in duration. For general analysis, let the output saturation voltages be denoted by $\pm V_{\text{sat}}$ and the threshold voltages by $\pm V_{\text{th}}$.

In the charging process, the initial and final voltages are $-V_{\text{th}}$ and V_{sat}, respectively. Applying the general formula for a first-order RC transient circuit, the voltage across the capacitor can be expressed as:

$$V_C(t) = V_{\text{sat}} - (V_{\text{th}} + V_{\text{sat}})e^{-t/\tau}, \tag{6.13}$$

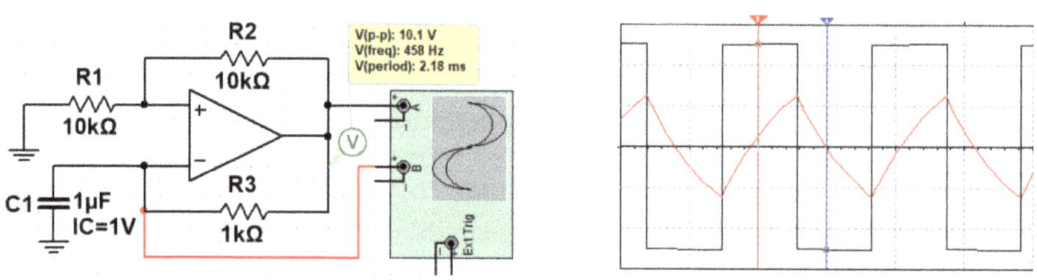

(a) (b)

Figure 6.39. Relaxation oscillator: (a) circuit, (b) waveform.

where the time constant is $\tau = R_3C_1$. This charging process ends when the voltage across the capacitor rises to $V_C = V_{th}$ at time t = T/2. Plugging this expression into equation (6.13) allows the period of oscillation to be determined:

$$T = 2\tau \ln \frac{V_{sat} + V_{th}}{V_{sat} - V_{th}} = 2\tau \ln(1 + 2R_1/R_2). \tag{6.14}$$

Using the component values shown in figure 6.39(a), the calculated period is 2.20 ms, which very closely matches the simulation result of 2.18 ms, confirming the validity of the analytical model.

> Relaxation oscillators operate based on the charging and discharging cycle of a capacitor, rather than sinusoidal resonance, making them inherently simple and highly robust. They are widely used in applications requiring clock pulses, blinking lights, pulse generators, and tone generation, especially where precise frequency tuning is less critical than in harmonic oscillators.

In the relaxation oscillator circuit, the Op-Amp can be effectively replaced by a pair of complementary metal–oxide–semiconductor (CMOS) inverters, forming a simple yet functional oscillator, as illustrated in the top-left section of figure 6.40. Four key nodes in this circuit are labeled *A* through *D*, and their corresponding waveforms are shown below the schematic. Nodes *C* and *D* are connected to the outputs of the two inverters. As expected, these outputs produce square waves oscillating between 0 and 5 V, with opposite logic levels—when node *C* is high, node *D* is low, and vice versa—demonstrating the inverter relationship.

On the other hand, the signals at nodes *A* and *B* are analog waveforms representing the charging and discharging behavior of the capacitor. Comparing

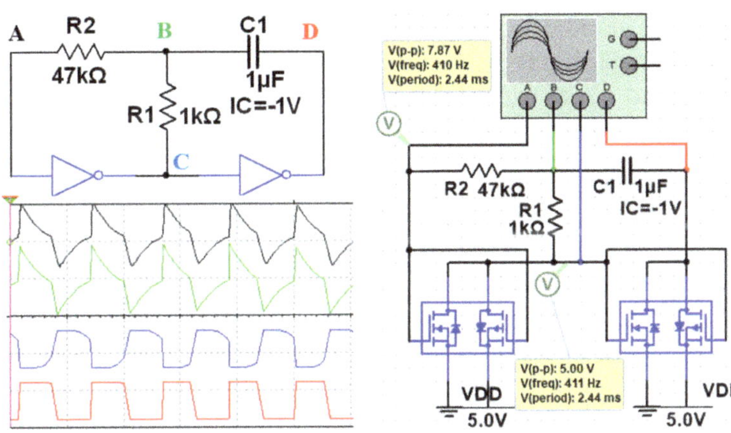

Figure 6.40. Relaxation oscillator with two inverters.

these two waveforms, we find that the signal at node A is just a copy of the signal at node B. Because the input of a CMOS inverter has extremely high input impedance —it is effectively an open circuit at low frequencies—virtually no current flows through the resistor R_2. This observation reveals that R_2 is functionally irrelevant to the behavior of this circuit.

The operation of this CMOS inverter-based relaxation oscillator can be analyzed as follows. Initially, assume that $V_A = V_B = 0$ V, $V_C = V_{DD}$, and $V_D = 0$ V. In this state, current flows upward through resistor R_1, charging capacitor C_1. As a result, the voltage at node B (and equivalently at node A) gradually increases over time. When this voltage reaches the threshold voltage of the inverter on the left side, the inverter switches state, causing its outputs to flip: $V_C = 0$ V and $V_D = V_{DD}$. As we know, the voltage across a capacitor cannot change instantly, so there is an overshoot of the voltage at node B beyond V_{DD}. The simulation result shown in the voltage probe on the left side indicates that $V_{A,\text{p-p}} = 7.87$ V.

This sudden change of the voltages at nodes B and C reverses the current through R_1, and the downward current through R_1 starts to discharge the capacitor. As the voltage at node B falls, it eventually crosses the inverter's threshold again, triggering the next transition of the inverter on the left side. This continuous charging and discharging cycle produces a stable square wave at nodes C and D, with the voltage at nodes A and B following charging and discharging waveforms centered around the threshold voltage.

However, the standard inverter model in Multisim is intended for pure digital logic simulation and does not include the necessary analog behavior, such as gradual transitions and the threshold voltage of transition, required for this oscillator to function correctly in simulation. Therefore, to accurately model the inverter behavior for this oscillator, two custom inverter subcircuits are constructed from discrete CMOS transistor pairs, as shown in the circuit diagram on the right side of figure 6.40. Each inverter consists of one n-channel MOSFET (NMOS) on the left and one p-channel MOSFET (PMOS) on the right, with their gates connected together to form the input and their drains connected together to form the output. The source of the NMOS is tied to ground, and the source of the PMOS is tied to V_{DD}, following the standard CMOS inverter configuration.

The period of oscillation is primarily determined by the RC time constant, set by the values of R_1 and C_1, as well as the switching thresholds of the CMOS inverters, which depend on the device parameters and supply voltage. This method of oscillator construction demonstrates how purely digital components can be arranged to produce analog timing behavior, bridging the gap between digital and analog design techniques.

A ring oscillator is a very simple yet fundamental circuit, consisting of an odd number of inverters connected in a closed loop, as shown in figure 6.41(a). The basic function of each inverter is straightforward—it outputs the opposite voltage level to that of its input. In this circuit, the high voltage level is $V_{DD} = 5$ V, and the low voltage level is ground (0 V). Therefore, when the input to an inverter is at ground, the output switches to 5 V, and when the input is at 5 V, the output drops to ground.

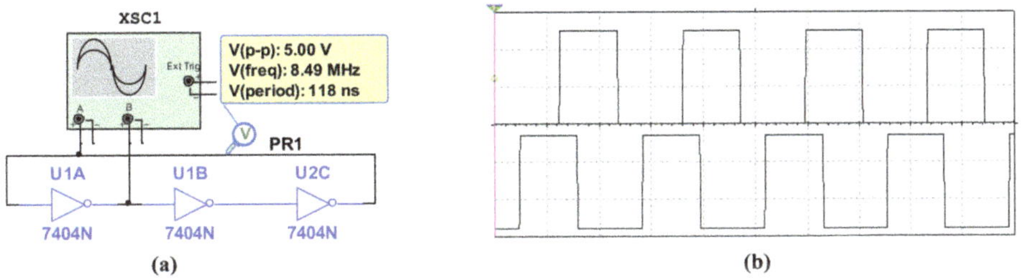

Figure 6.41. Ring oscillator: (a) circuit, (b) waveform.

However, this change does not occur instantaneously. Each inverter has a finite propagation delay, primarily due to the intrinsic capacitance and the finite drive current of the transistors inside. This delay is clearly observable in the waveform diagram shown in figure 6.41(b), where the top waveform is the input of the first inverter (U1A), and the bottom waveform is the output of this inverter. Closer observation reveals that the delay associated with the rising edge (pull-up delay) is typically different from that of the falling edge (pull-down delay).

In CMOS technology, an inverter consists of a complementary pair of transistors: an NMOS and a PMOS. The NMOS transistor conducts when the input is high, pulling the output voltage down to ground, while the PMOS conducts when the input is low, pulling the output up to V_{DD}. A key characteristic of CMOS design is that the NMOS is inherently faster than the PMOS. This is because the mobility of electrons in the NMOS channel is significantly higher than the mobility of holes in the PMOS channel, resulting in faster switching for pull-down transitions compared to pull-up transitions. This intrinsic asymmetry in charge carrier mobility leads to unequal propagation delays for rising and falling edges.

The device model of inverters in Multisim is intended for use with digital circuits, and it is greatly simplified. In this model, the output is either logical '1' (V_{DD}) or logical '0' (0 V), and the transition delays are different: the pull-up delay is 22 ns, while the pull-down delay is 15 ns. Therefore, the total delay in one period is 37 ns. The oscillation period of a ring oscillator is directly determined by the cumulative delays of the individual inverters. Specifically, for a ring oscillator with three inverters, each contributing a combined pull-up and pull-down delay of approximately 37 ns, the total delay sums to around 111 ns for one full oscillation cycle. This calculation aligns closely with the simulated period of the square wave, measured at 118 ns in figure 6.41(a), confirming the accuracy of the delay-based oscillation model.

It is very easy to construct a ring oscillator circuit in the lab; figure 6.42 shows waveforms captured from ring oscillators built using inverters in an SN7404N chip, which is based on TTL technology. The measured parameters indicate that the oscillation periods are approximately 37.6 ns for the circuit with three inverters in figures 6.42(a) and 65.0 ns for the circuit with five inverters in figure 6.42(b). As discussed earlier, the period is expected to be approximately proportional to the number of inverters in the circuit, since each stage contributes a fixed propagation

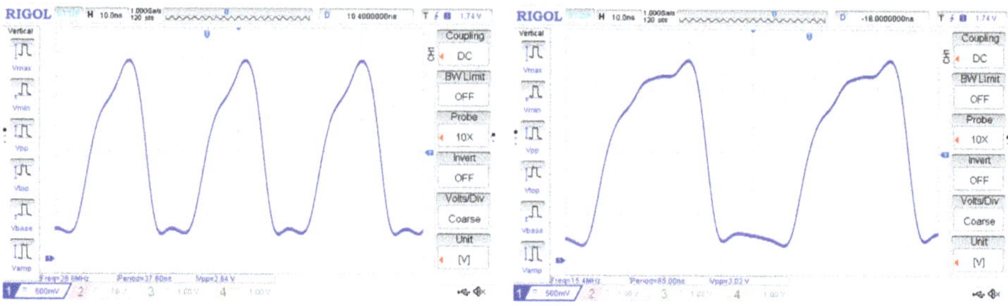

(a) **(b)**

Figure 6.42. Experimental results obtained from ring oscillators: (a) three inverters, (b) five inverters.

delay. The experimental results shown here clearly verify this linear relationship: $T_1/T_2 = 0.578 \approx 3/5$.

In addition to the period, the waveforms reveal other important characteristics. Notably, there is a distinct asymmetry between the rising and falling edges. The falling edge is much steeper, corresponding to the faster pull-down action of the *n*-type transistors, whereas the rising edge appears slower and visibly deformed due to the slower pull-up response of the *p*-type transistors, whose carrier mobility is lower. To achieve cleaner waveforms suitable for precise timing or further digital processing, additional waveform conditioning—such as passing the signal through buffer stages or Schmitt triggers—may be necessary to sharpen the edges and restore signal integrity.

The ring oscillator is a powerful circuit clearly demonstrates how oscillation can emerge solely from signal inversion combined with propagation delay in a feedback loop, without the need for inductors, capacitors, or any form of resonant energy storage. Despite its simplicity, the ring oscillator plays a crucial role in ICs, particularly in clock generation, on-chip timing references, frequency synthesis, and random number generation, thanks to its ease of implementation and full compatibility with standard CMOS processes.

> From a certain point of view, oscillation represents a fundamental form of existence in nature. In a ring oscillator, the presence of an odd number of inverters creates a logical contradiction that prevents the circuit from settling into a stable state. Instead, the circuit is forced to perpetually toggle between high and low voltage levels, resulting in continuous oscillation. Figuratively speaking, this inability to reach a stable solution forces the circuit's behavior to extend into the time domain, manifesting as oscillatory motion.

In contrast, if the number of inverters in a feedback loop is even, the circuit can settle into one of two possible stable states rather than oscillating. This bistable behavior forms the foundation of digital memory elements and belongs to the category of bistable multivibrators. Such circuits maintain their output

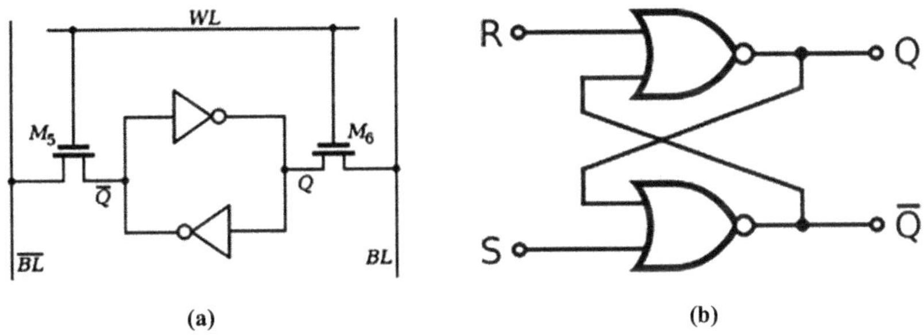

Figure 6.43. Ring with two logic devices: (a) SRAM. This [SRAM Cell Inverter Loop] image has been obtained by the author from the Wikimedia website where it was made available by [Crystallizedcarbon] under a CC BY-SA 4.0 licence. It is included within this book on that basis. It is attributed to [Crystallizedcarbon]. (b) SR latch. Created with GPT-4.0, OpenAI.

indefinitely until an external input forces them to change state. A typical example is the static random-access memory (SRAM) cell—widely used as cache memory in modern CPUs—which is constructed from a cross-coupled pair of inverters, as shown in figure 6.43(a). This configuration can reliably store one bit of information by holding either a logic '1' or '0'. Similar bistable multivibrator circuits form fundamental digital building blocks such as latches and flip-flops, which are essential for data storage and sequential logic operations in digital systems.

Based on the SRAM circuit, fundamental sequential logic components can be constructed. For example, if the inverters are replaced by NOR gates or NAND gates, each with an additional control input, the circuit becomes an SR (set-reset) latch, which is shown in figure 6.43(b). To see the similarity between these two circuits, the bottom NOR gate in this diagram can be flipped horizontally, and then the cross-coupled connections become visually disentangled, revealing a topology that closely resembles the inverter ring structure of an SRAM cell. This similarity highlights a key conceptual connection: both circuits rely on positive feedback between two logic components to maintain a stable state indefinitely.

> Compared with the SRAM cell, there are external control inputs in the SR latch that directly manage state transitions—allowing the output to be explicitly *set* (logic '1') or *reset* (logic '0'). The SR latch represents one of the simplest forms of sequential logic components, serving as the foundation for more complex components such as gated latches, flip-flops, and registers widely used in digital electronics.

In addition to astable and bistable multivibrators, the last type is the monostable multivibrator. It is also known as a one-shot circuit and has one stable state and one quasi-stable state. Under normal conditions, the circuit remains in its stable state. When triggered by an external pulse, it temporarily switches to the quasi-stable state for a predetermined duration before automatically returning to the

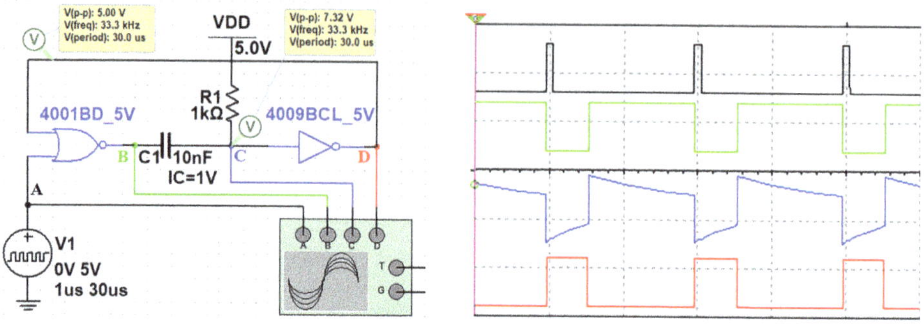

(a) (b)

Figure 6.44. Monostable multivibrator: (a) circuit, (b) waveform.

stable state. Monostable multivibrators are widely used for generating precise timing pulses, pulse width modulation, delay circuits, and debouncing switches.

Figure 6.44(a) shows a simple monostable multivibrator circuit, which consists of an inverter, a NOR gate, and an RC timing network. In the stable state, the output at node *D* is held at 0 V. When the input at node *A* is also at 0 V, the NOR gate output at node *B* is driven high to 5 V. After a sufficient time, node *C* eventually reaches 5 V as well. At this point, no current flows through resistor R_1, and the system remains in a stable state. This condition persists until an external trigger pulse is applied to node *A*, which initiates the transition to the quasi-stable state, during which the output briefly switches before returning to the stable state.

The signal source located at the lower-left corner of the circuit provides periodic trigger pulses, as illustrated in figure 6.44(b). In this diagram, the four traces represent the voltages at nodes A–D, respectively. When a pulse arrives and drives the input at node A to 5 V, the output of the NOR gate (node B) drops to 0 V, since a NOR gate outputs a low (0 V) whenever any of its inputs are high. This sudden drop at node *B* forces node *C* to drop simultaneously because the voltage across a capacitor cannot change instantaneously. As a result, the inverter detects the low voltage at node *C* and switches its output (node *D*) to 5 V, generating the output pulse.

The inverter output is directly fed back to the other input of the NOR gate, creating a latch mechanism that holds node *B* at 0 V even after the input pulse from the signal source returns to 0 V. During this quasi-stable period, a current begins flowing downward through resistor R1, charging the capacitor. Consequently, the voltage at node *C* gradually rises toward 5 V following an exponential curve governed by the RC time constant. The values of resistor R1 and capacitor C1, as well as the threshold voltage of the inverter, directly determine the duration of the output pulse.

When the voltage at node *C* crosses the inverter's threshold voltage, the inverter output at node *D* switches back to 0 V. This transition removes the latch condition, allowing the NOR gate output at node *B* to return to 5 V. The circuit returns to its stable state, ready to be triggered by the next pulse from the signal source.

In summary, multivibrators are fundamental circuits that generate non-sinusoidal waveforms, primarily used in timing, pulse generation, and switching applications. They are classified into three types: astable, monostable, and bistable. An astable multivibrator has no stable state and continuously oscillates, generating square waves, making it suitable for clock generation. A monostable multivibrator has one stable state and produces a single output pulse in response to an external trigger, functioning as a one-shot pulse generator. A bistable multivibrator has two stable states and can store one bit of information, forming the foundation of memory elements such as flip-flops and SRAM cells. These circuits are essential in both analog and digital systems for control, timing, and data storage functions.

A monostable multivibrator shares conceptual similarities with the behavior of biological neurons. Both systems exhibit an excitable response to external stimuli. In a neuron, when the membrane potential reaches a certain threshold due to an external stimulus, the neuron generates an action potential—a rapid transient spike in voltage. After firing, the neuron returns to its resting state, ready to respond to the next stimulus.

Similarly, a monostable multivibrator remains in a stable state under normal conditions. When it receives a trigger pulse, it switches to a temporary quasi-stable state, producing a fixed-duration output pulse, before returning to its stable state automatically. This transient response mirrors the all-or-nothing firing behavior of neurons. In essence, both systems act as threshold-triggered pulse generators—neurons in biological systems and monostable multivibrators in electronic systems—highlighting how certain principles of electronics parallel biological information processing.

Bibliography

Semiconductor electronic devices

1. Neamen D A 2011 *Semiconductor Physics and Devices: Basic Principles*, 4th edn (New York: McGraw-Hill)
2. Streetman B G and Banerjee S K 2014 *Solid State Electronic Devices*, 7th edn (Boston, MA: Pearson)
3. Sze S M, Li Y and Ng K K 2021 *Physics of Semiconductor Devices*, 4th edn (Hoboken, NJ: Wiley)
4. Tsividis Y and McAndrew C 2010 *Operation and Modeling of the MOS Transistor*, 3rd edn (Oxford: Oxford University Press)
5. Hu C 2009 *Modern Semiconductor Devices for Integrated Circuits* (Boston, MA: Pearson)

Electronic circuits

1. Irwin J D and Nelms R M 2021 *Engineering Circuit Analysis*, 12th edn (Hoboken, NJ: Wiley)
2. Alexander C and Sadiku M 2016 *Fundamentals of Electric Circuits*, 6th edn (New York: McGraw-Hill)
3. Sedra A S, Smith K C, Carusone T C and Gaudet V 2019 *Microelectronic Circuits*, 8th edn (Oxford: Oxford University Press)
4. Neamen D A 2009 *Microelectronics Circuit Analysis and Design*, 4th edn (New York: McGraw-Hill)
5. Razavi B 2021 *Fundamentals of Microelectronics*, 3rd edn (Hoboken, NJ: Wiley)
6. Horowitz P and Hill W 2015 *The Art of Electronics*, 3rd edn (Cambridge: Cambridge University Press)
7. Franco S 2014 *Design with Operational Amplifiers and Analog Integrated Circuits*, 4th edn (New York: McGraw-Hill)

doi:10.1088/978-0-7503-5512-4ch7

8. Young P H 2003 *Electronic Communication Techniques*, 5th edn (Upper Saddle River, NJ: Prentice-Hall)
9. Gonzalez G 2008 *Foundations of Oscillator Circuit Design* (Norwood, MA: Artech House)
10. Lee T H 2003 *The Design of CMOS Radio-Frequency Integrated Circuits*, 2nd edn (Cambridge: Cambridge University Press)
11. Razavi B 2016 *Design of Analog CMOS Integrated Circuits*, 2nd edn (New York: McGraw-Hill)
12. Carusone T C, Johns D A, and Martin K W 2011 *Analog Integrated Circuit Design*, 2nd edn (Hoboken, NJ: Wiley)
13. Gray P R, Hurst P J, Lewis S H and Meyer R G 2024 *Analysis and Design of Analog Integrated Circuits*, 6th edn (Hoboken, NJ: Wiley)
14. Allen P E and Holberg D R 2011 *CMOS Analog Circuit Design*, 3rd edn (Oxford: Oxford University Press)
15. Weste N and Harris D 2010 *CMOS VLSI Design: A Circuits and Systems Perspective*, 4th edn (Boston, MA: Pearson)
16. Baker R J 2019 *CMOS Circuit Design, Layout, and Simulation*, 4th edn (Piscataway, NJ: Wiley-IEEE Press)
17. Plummer J D and Griffin P B 2024 *Integrated Circuit Fabrication: Science and Technology* (Cambridge: Cambridge University Press)

www.ingramcontent.com/pod-product-compliance
Ingram Content Group UK Ltd.
Pitfield, Milton Keynes, MK11 3LW, UK
UKHW051856100926
8556IPUK00005B/646

9 780750 355100